Beginner's Guide to Embedded C Programming

Using the PIC® Microcontroller and the HI-TECH PICC-Lite™ C Compiler

By Chuck Hellebuyck

Published by Electronic Products.

The publisher offers special discounts on bulk orders of this book.

For information contact:

Electronic Products
P.O. Box 951
Milford, MI 48381
www.elproducts.com
chuck@elproducts.com

Printed in United States of America
Cover design by Rich Scherlitz

Table of Contents

Introduction

I've been programming microcontrollers in assembly language and high level BASIC language for many years. My experience with programming in general dates back to the early personal computer (PC) days when QBasic was the popular choice. I was content with these programming skills so I never seriously tried to learn the C language even though is was clearly becoming more popular. C is commonly used for developing PC applications and also became the standard for professionals that write code for microcontrollers, also known as embedded C programming. I decided it was time to get serious and learn C. Specifically embedded C for my favorite family of embedded microcontrollers called the PIC® microcontroller (MCU) from Microchip Technology Incorporated. I tried to find a good class or even a good book for the beginner that focused on this topic. After purchasing several books and even attending a class on the topic, I still didn't find that entry level guide I was looking for that included the basics plus actual hardware projects. I had taught myself PICBASIC this way so I decided I would have to teach myself C with the idea that I would create the book I wanted to buy. I took a lot of notes and a big portion of those notes are between the pages of this book.

Even though I have been very successful programming these microcontrollers in assembly and BASIC I had to find out why the C language was the professional programmer's language of choice for programming embedded applications. It was a bit of a mystery to me why this language was the first choice since I saw it as kind of cryptic. I had heard from many professional programmers that the C language is great and is easier to learn than most people think. I wanted to end my confusion over this language so I decided to start simple and try and make an LED light up using the freeware version of the PICC-Lite™ C compiler from HI-TECH Corporation. I also discovered that this sample version of their professional PICC compiler was perfect to accompany this book. You can download it for free from www.htsoft.com and the best part is the sample version isn't time limited so you can use it forever as long as the memory limits they put on the select parts it supports meet your needs.

At first it was amazing how difficult getting my first program to work became. I had all kinds of errors that baffled me. What I found was I needed to change the way I thought about programming. I had to dive a little deeper into writing the compiler functions that BASIC compilers tend to do automatically with pre-written commands and also stop trying to control every bit that assembly language requires. I had to give C a fair chance to enter my confused brain. What I found is the C language falls in-between a true high level language like BASIC and a low level programming language like assembly but it does offer a lot of options to the programmer.

I then thought it would help to study various sample programs I could find on the internet but I didn't find that method very helpful. In fact I saw so many different examples of doing the same thing I began to become even more confused. My mistake was making the assumption that C code would be common from program to program or at least follow some kind of similar approach. This assumption came from the reputation that C code is universal and can be easily converted from one platform to another. It turns out if you are talking about programming PC's this is true because the operating system creates a common platform to build on. All this falls apart though when you are talking about embedded programming. Microcontrollers from different companies each have their own structure.

The C language can easily adjust to these different choices because of its versatility but this versatility also adds to the confusion. The various C programs I reviewed were written on different C compilers and they each had various structural differences. You see the C language has evolved over the years and many short cuts to do the same thing have been added for the experienced C user and also to make the C language work with the embedded world rather than just the PC world. This evolution has expanded the options the user must understand but this also adds to the confusion the beginner will experience.

While I went through my learning process I took lots of notes. This book is a partial summary of my notes and is intended to teach you how to program in embedded C without the large learning curve. The key point to understand is this book is focused on the beginner interested in embedded C programming not general C programming. It also doesn't go deep into all the options C programming offers. Pointers are a topic I'm saving for a follow-up book along with other topics I consider advanced.

This book will take you through some of the very basics of the C language and then teach you programming by example with a very low cost Starter Kit available from Microchip Technology. I chose this platform because it's a complete package with microcontroller included, is inexpensive and also includes the freeware version of the HI-TECH PICC-Lite compiler on the kit's CD. This is everything you need to do the projects in this book. All the project software in this book was written by me and tested by me and had many mistakes created by me while doing it. By using that method I can pass along what to watch out for and know what you need to learn to get through it faster than I did. Isn't that the point of a book anyway?

I hope in the end you find this book worth keeping as a reference as you eventually pass me up and become a far better C programmer than I will ever be. At that point I can say I was a successful teacher and this book was worth the price. Lets get started learning how to program in embedded C.

Chapter 1 – What is Embedded C Programming

I frankly don't know the complete history of C and thankfully you don't need to know it to actually use this programming language. It does help though to know some history of embedded programming to understand where it all began. When computers were just starting out they were actually embedded machines that only ran one program at a time. The computer was one of the first truly successful applications of the microprocessor. I've often read that the Microsoft founders created a traffic counter box out of a microprocessor before they created the operating system we now know as MS-DOS or Microsoft Disk Operating System. I've also read that the founders of Apple Computer also created small projects from microprocessors before creating the Apple 1 computer kit. This is how they learned.

Personal Computers took off quickly and eventually added operating systems and then graphical interfaces such as Windows to make using the computer easier. All those applications that run the PC had to be created in some language and the C language created by Dennis M. Ritchie of Bell Labs in the early 1970's for mainframe computers became a popular choice for desktop applications. Eventually the microprocessor manufacturing technology began to get smaller and more circuits could be put inside one single integrated circuit (IC). The microcontroller (as opposed to microprocessor) was created that contained all the key components to run a single application within the IC. That programmed IC or chip could be built into a product so the end user didn't even have to know how to run a computer to use the application. This became known as an embedded application as the software and hardware were embedded into the product or design. Just as a PC motherboard has a microprocessor, ROM (hard drive), RAM and input/output circuitry connected through the BIOS, the microcontroller has all that, albeit smaller, inside a single IC.

The early microcontrollers were programmed in assembly language but there was a definite desire to program in a higher level language so some of the techniques used in creating PC applications could be used in the embedded world of microcontrollers. As the memory size and capabilities of microcontrollers increased so did the desire for languages that could simplify the task of programming the embedded application. Languages such as BASIC or PASCAL were used by some but clearly the most popular was the C language.

There are some who claim that C is a high level language. I tend to disagree. Assembly language is the lowest form of language for embedded microcontrollers and languages like BASIC and PASCAL are easier to read and understand than most other languages so I consider these high level languages. C language falls somewhere in between. I say this because a BASIC compiler for the PIC Microcontroller (MCU), such as PIC BASIC PRO that my first book "Programming PIC Microcontrollers in PicBasic" was written about, makes writing software very simple. Many functions are already created so a single command can be used to drive a Liquid Crystal Display (LCD) or send a serial RS232 style message. The C language requires you to write all those steps yourself unless you can find somebody to share code they wrote to perform that function or find a routine in the library of functions included with the C compiler.

In fact the word "function" becomes very important in the C language world. In another language such as BASIC you might write a chunk of code that performs a function over and over again. To save time and memory space you might turn that chunk of code into a subroutine that you might call over and over again throughout your main loop. In the C language a subroutine is known as a "function" and most C programs are just a bunch of functions spliced together with some other C language statements to form the complete program. Some of these pre-written functions are included with the compiler and called a library.

Another confusing term you might hear that I've already used is the term "statement". In BASIC or other languages what is referred to as a command is called a "statement" in C. Statement examples are *do-while*, and *for-loop*. The projects later in this book will give simple examples of these. These statements will be used to build the functions and the functions will be glued together using more statements. The great advantage I see to the C language is the list of statements is very short so it's quite easy to remember them. You can create all kinds of custom programs much quicker than assembly but still have the freedom that BASIC or PASCAL may limit.

This book is written to help you, the reader, learn how to get started programming Microchip PIC® microcontroller's (MCU) with the C language. The PIC Microcontrollers or MCU's are very popular microcontrollers and are used in many embedded applications. One of the more popular C compilers for the PIC MCU's is the HI-TECH PICC compiler that is also offered in a sample version called the HI-TECH PICC-Lite. This sample version of the compiler is not a typical version that expires in 30 days. Instead it just limits how much of the PIC MCU's memory you can use. All the examples I present in this book can be accomplished with the lite version of the compiler so I find it a great companion to this book.

Microchip offers over 300 different PIC MCU's at the time that I write this and more are on the way. To simplify the teaching process though I wanted to choose one that was easy to get, low cost, had lots of features and was supported by the PICC-Lite compiler. I easily found it already included in the starter kit offered by Microchip under the part number PIC16F690. Since you also need the tools to program the microcontroller and test your code I found a one stop source for everything we need to learn C programming; the PICkit™2 Starter Kit from Microchip.

Microchip offers the PICkit 2 Starter Kit for $49.95 and you can buy it from various suppliers including Microchip's own website; microchipdirect.com. This starter kit includes a USB powered PIC MCU programmer, 20 pin socket development board with LEDs, potentiometer, momentary switch and expansion area plus a PIC16F690 microcontroller in the package. In addition to this, the package includes a CD with Microchip's MPLAB® Integrated Development Environment (IDE) for writing the C code on a PC and the CD also has the PICC-Lite compiler included so you can load that on the PC as well. I wrote all the projects in this book using this setup so for the price of this starter kit in addition to the price you paid for this book, you have everything you need to learn the basics of how to program in embedded C.

Before we create the first project though I wanted to cover the basics of a C program and the next few chapters takes you through that. By no means is this a complete C language summary but rather just the basics for beginners to offer enough of the key information you need to get your first programs running. Beyond that, this book will also step you through setting up the

PICkit 2 Starter Package and how to use it within the Microchip MPLAB design environment most professional PIC programmer's use. I didn't want to leave out the users who have never programmed a microcontroller directly though I assume the reader has at least some electronics background.

Now lets begin with the what exactly a C program looks like.

Chapter 2 – C Program Structure

To really understand a C program I decided to start from the top and show a very simple main loop of code. Listing 1 shows the main loop of code that I will use as the starting point for the projects in this book.

```
/*************************************************************
File:                    template.c

Description: This file contains the starting template for a C program
written in HI-TECH PICC Lite and PIC16F690
*************************************************************

Created By:  Chuck Hellebuyck 10/18/06

Versions:  1.0

*************************************************************/

#include <pic.h>      // Include HITECH CC header file

/*Internal clock, Watchdog off, MCLR off, Code Unprotected,
Power Up Disabled, Brown Out Reset Disabled */

__CONFIG (INTIO & WDTDIS & PWRTDIS & MCLRDIS & BORDIS &
UNPROTECT );


main()
{
// your program code is entered here
}  //end main
```

Listing 1 – C Program Template

At first glance the program in Listing 1 may look like a bunch of cryptic characters and not really a program at all. In a way that is correct. This program doesn't do anything except form the basis of a C program. Surprisingly though it will compile without errors and can be considered a complete C program. I created this just to show the template for starting an embedded C program. Let me step through this so you understand how an embedded C program is structured.

At the top is a comment block where the description of the program is established. Notice how the very first character is a "/" forward slash character followed by a "*" star character. This combination "/*" indicates to the compiler that the characters that follow are all comments. In fact those comments can continue for several lines. The only thing that stops the C compiler from reading every character as a comment is another "/" and "*" only this time in the opposite order "*/". This is one way of creating comments across multiple lines. Start the comment section with a "/*" and end with a "*/". The C compiler will not recognize any carriage return characters between these comment characters so the header block can spread across multiple lines and be treated as a single line. Using the "/*" "*/" combination makes it easy to create header blocks in your code.

Notice that I also used stars to create separation lines in the header block. This can actually create an error to watch out for. If the next character following one of those stars in my header block is a "/" character then the compiler will read that as the end of the comment line and the rest of the comments will be treated as programming control statements. This can be hard to find when debugging so use the stars carefully. In fact you might want to use an underscore or some other line separating character. I used the stars to help point out a common beginner error (at least it was for me).

```
#include <pic.h>      // Include HITECH CC header file
```

The next line is a pre-processing directive. In other words it is executed before the program gets compiled. The "#include" is the directive and tells the pre-processor to include the contents of a file before compiling the code. In this example the file to include is the "pic.h" file. This is known as a header file and the ".h" indicates this. The pic.h file is included with the HI-

TECH PICC-Lite compiler and contains a bunch of special nicknames and program setup features. The arrow brackets around the file name indicate that the file is in the same directory as the compiler. I'll explain more about this "pic.h" file later as it is very important to programming with PICC-Lite. Even if you use a different C compiler, it will have some kind of header file to include for the compiler you are using.

After the "pic.h" file is listed, you will notice a comment stating "Include HITECH CC header file" but you don't see a "/*" or "*/". Instead you will see a "//" double forward slash. This is another way of creating a comment but the difference is it only lasts until the end of the line. This is one of the rare times that the C compiler recognizes an end of line without some special character. You can create a program header by placing "//" in front of the every line instead of using the "/*" "*/" method but that is not typical.

```
/*Internal clock, Watchdog off, MCLR off, Code Unprotected,
Power Up Disabled, Brown Out Reset Disabled */

__CONFIG (INTIO & WDTDIS & PWRTDIS & MCLRDIS & BORDIS
& UNPROTECT );
```

The next line in the template is another comment line that starts with the slash-star. The comment line describes some features of the microcontroller. These are just comments to make it easier to understand the configuration setup. The real control is in the next line that starts with __CONFIG. The double underscore followed by the word CONFIG is a directive to the compiler that the characters that follow are the setup fuses or configuration bits that will be needed when actually programming the microcontroller. This is an embedded C line and you won't find this in a C program written for a PC. This line will be different for each microcontroller you use and will be different if you use something other than a PIC Microcontroller. I will cover these in more detail later.

Some more experienced users may point out that this line is not really necessary as you can also set these in your programmer's software interface when you actually program the part but it is a best practice to include them because then the configuration setup stays with the code. This prevents the wrong configuration setup being programmed into the microcontroller by someone who didn't know the original setup. Some people just place the configuration description in the title block for reference but I find including this single line in every program is a simple and safe way to get the configuration correct.

Notice how the __CONFIG statement takes two lines to fit it all in. Just as the comment header block could span more than one line, programming control lines can also span more than one line. To the C compiler the second line looks like a continuation of the first line. Therefore something has to tell it when to stop or when the command line is finished or else the whole program could look line one continuous line. The ";" semicolon at the end of the second line is that indicator. In fact it is one of the biggest sources of errors for the beginner especially if you are converting from another compiler such as BASIC. Every command line in your program has to end with a semicolon but there are exceptions. Some command lines are designed to spread across several lines so the semicolon may not show up until later but it will be needed. I will cover this more in the project examples in future chapters as I describe the different statements.

A requirement of every C program is to have a main loop and that loop typically will start with the function name "main ()" followed by a left and right parenthesis. I'll explain more about functions later in the book but to put it simply a function is a subroutine that you create. The main function is the only required function but you will find that most of your programs will contain multiple functions and these functions will be called from the main function. The main function is where the program will start operating once it has been programmed into the microcontroller. When the compiler creates the binary file to be programmed into the microcontroller it has to know where you want to start executing code from. The main function is that place. Some compilers allow you to set that location in the microcontrollers memory thru a file known as the "linker script". The PICC-Lite compiler performs this linker function somewhat in the background so this makes it a bit easier to use especially for the beginner.

```
main()
{
// your program code is entered here
```

```
} //end main
```

Below the word main will follow two curly brackets. One left and one right. Between these brackets will be your code. I put a comment line to show you where your program code will go but future project examples will explain it better. The main point I wanted to make is this simple section of code will compile without errors. It will make a program file that doesn't really do anything but it will compile and create a binary file that can be programmed into a microcontroller. In fact, there is a unique difference between this structure for an embedded application and one for a PC. This same setup will compile for a PC but an implied GOTO statement will be inserted at the end of the PC version so the program will keep it looping back to the top.

In the embedded world that doesn't happen. This program will run though once and then stop. If the internal watchdog circuit is enabled on the microcontroller then the part will reset itself and only then will it run the code again. This is why you always want to have some type of loop control statement encompassing your overall program. This is easily done with a *while*(1) statement at the top or other looping options I'll explain in the following chapters.

This method of looping is no different than a BASIC language program that ends with a GOTO MAIN command line to loop the program back to the top. I just wanted to point this out in case you try to compare a simple C program written for a PC vs. one written for an embedded microcontroller application. It's those little things that can trip you up when just getting started.

One thing this program doesn't show is a blank line after the final line of characters. With the HI-TECH compiler and other compilers I've actually gotten errors because I didn't have a blank line inserted at the end. I've also found some C compilers don't require that. It was a frustrating bug to find as it prevented my program from compiling without errors. Just keep that in mind as you begin to write your own embedded C programs.

Linking C Files

The simple example I've shown of a main loop of code would create a single C file typically called main.c but it could be called anything just as long as it has the ".c" suffix to indicate it is a C program file. Some of the other information required for that file may be placed in a separate header file or .h file. It may even share the same name with a ".h" suffix such as main.h. This is actually a common practice and what makes the C language so useful. You see you can have a separate ".c" file for every part of your overall program. You can have an lcd.c file for driving a liquid crystal display or led.c for driving a bank of light emitting diodes (LED). Each of those ".c" files may have variables or constants defined in a separate ".h" file associated with each ".c" file. This is very common and makes it easy to pass on these various routines from program to program and easily shared between those creating their own C language program. This will save you a lot of time if you can find someone who has already written an LCD program so all you have to do is include it with your program.

At this point you are probably wondering how all these different files get combined into one binary file that can be programmed into the microcontroller. The answer is the linker script. Every C compiler has a special program called the linker script that takes all the various ".c" files and ".h" files and combines them into one complete program prior to compiling and then assembling the program into a binary file. That binary file for Microchip microcontrollers is always in a hexadecimal format with the suffix ".hex". Each C compiler handles linking slightly different and the HI-TECH PICC-Lite compiler used in this book kind of hides that step from the user to make it easier to use. You can get access to the linker and make some modifications but that is really for the advanced user so I don't suggest you even worry about it. Just know that when you have several ".c" and ".h" files, they will be linked into one file. In fact Project 6 in Chapter 13 shows an example of how to link files.

Header Files

Defined variables and constants along with other setup information are often placed in a header file or ".h" file. The same way the HI-TECH PICC-Lite Compiler requires you to include the "pic.h" file at the top of the program; you can add your own header files at the top of your program. The advantage is the ease of finding and changing variables or setup information in the header without having to modify the ".c" file. The "pic.h" file contains lots of special characters and definitions the HI-TECH PICC-Lite Compiler uses to compile your code for the Microchip microcontrollers. Each person will use header files in their own way so there really isn't a standard or "right way" to do this from what I've found. It's just an added feature C language allows you to use.

Other than the "pic.h" header though, you don't have to include any header file. You can put all the setup, variables and constants in your main.c file and never have separate ".c" files for all the various functions you want your program to perform. That will make for a very large ".c" file but if you've programmed in Embedded Basic or Assembly you may already use that method. Hopefully you now understand that C is really similar in structure but offers other options not often used in other language compilers.

Variables and Constants are another area where the C language begins to offer differences not seen in assembly or compiled Basic. I'll cover that next.

Chapter 3 – Variables, Constants and Arrays

Almost every program you write will have some kind of variable or constant value included in the various functions the program performs. In the embedded world of C programming that means using the Random Access Memory (RAM) area of the microcontroller. If you are using a chip with a very small amount of RAM then this may limit the amount of variables you can use in your program. One of the advantages of C that I found when compared to Assembly or BASIC is the re-use of RAM space. The C language uses two kinds of variables; global and local.

Global variables have a fixed location in RAM and can be accessed by different functions. This means the value may get altered by one function and then later get changed by a totally different function. This is what makes it global. The C language also allows multiple small C programs to be combined or linked into one file so the compiler will automatically recognize the multiple use of a variable and make the necessary connections in software so each of the individual ".c" files can point to the proper RAM space containing the variable.

There are other times when you only need to use a variable for a short snippet of code and after that the value can be erased. This is known as a local variable and the RAM space used for that variable will be used over and over by several different functions within the same ".c" file. For example, in the snippet of code below the variable "x" is created just before entering a *for-loop*.

```
while(1==1)                     //loop forever
      {
      int x;
      for(x=1; x<9; x=x*2)
            {
            PORTC = x;              // Turn on next LED
            Pause(delay);           // Delay for .5 seconds
            }// End For

      }//End while
```

After the *for-loop* is complete, we really don't care about the value stored in the variable "x". If later on in the program another routine uses a *for-loop*, it can create the variable "x" again or even a variable of a different name but the compiler will recognize these as local variables that only need to be used temporarily so the same RAM space may be used for both these variables plus others if they are created. This saves RAM usage and is an advantage to C programming.

There are times when you may want to create a variable but have it keep its value after the function has completed its task. This may be a variable that stores how many times a switch has been pressed or how many times an interrupt has occurred. In this case you have to declare the variable as a "static" variable and it will receive a fixed RAM location. The term "static" is known as a storage class.

Storage Classes

Each compiler may handle this slightly differently so I will cover the PICC-Lite compiler here. PICC-Lite breaks variables down into two storage classes called "auto" and "static". If you declare a variable and don't specify the storage class then it will be automatically declared an "auto" variable. This means it will share RAM space with other "auto" variables within that same ".c" file. As I mentioned earlier, a variable that you want to keep its contents and have its own RAM address is given the "static" storage class. Though this may act like a local variable for the function using it, because it has a fixed memory location, it can also be accessed from other functions and could be considered a global variable. You have to really know how you want the variable to be used.

If you were previously programming in Assembly or a BASIC compiler then you were probably creating fixed RAM locations for all your variables. Therefore your variables were all global and none of them were local. Therefore the distinction between local and global variables can add a layer of confusion to the C language beginner. Nothing stops you from declaring all variables "static" and each with a unique name so they get their own RAM location but you would have to declare all variables at the beginning of your program and this will not use RAM space efficiently which is one of the advantages to programming in C. It's best to study the manual for the compiler you use to fully understand the options you have to use RAM efficiently.

Here are two examples of declaring a byte size variable named "count" as an "auto" storage class. They are both identical but the second one uses the PICC-Lite default declaration of "auto".

```
auto char count;  // Create variable "count" a 8-bit value.
     char count;  // This is another way of doing the same thing.
```

If you need the variable count to be accessible to other areas of the program or want it to have its own fixed location in RAM then this is how it would look.

```
static char count; // Create variable "count" as 8-bit fixed RAM location
```

Data Types

Notice that I stated the previous variable "count" was a byte sized variable yet the term "byte" is not stated anywhere in the declaration. This is another area of confusion for the beginner moving from another compiler such as PICBASIC. The C language has several different data type declarations that determine the size or amount of RAM space the variable will take up. The information in Table 1 shows the list of data types.

Type	Size (bits)	Arithmetic Type
bit	1	Unsigned integer
char	8	Signed integer
unsigned char	8	Unsigned integer
short	16	Signed integer
unsigned short	16	Unsigned integer
int	16	Signed integer
unsigned int	16	Unsigned integer
long	32	Signed integer
unsigned long	32	Unsigned integer
float	24	Real number
double	24 or 32	Real number

Table 1: Data Types

You can see that variable types can range from bit size all the way up to 32 bits wide. This table is not completely standard across all C compilers. In other compilers an “int” may be 8 bits wide. This isn’t typical but it can add some confusion when looking at C code from one compiler and comparing to another compiler. The values in Table 1 are the most typical sizes though. The variable declaration also has another characteristic defined. The C compiler also allows you to declare the variable as signed or unsigned. This sets the range of the variable. For example, an “unsigned char” variable has a range of 0 to 255 decimal or 0x00 to 0xFF hexadecimal which fits within 8 bits. A signed char uses those same 8 bits but the most significant bit is used as a sign indicator. If that bit is set or a one then the value is a negative number. If that bit is cleared or zero then the value is a positive number. Therefore you only have seven bits left to define the actual value. This gives you a range of -127 to +127 (actually it gives you -128 to +127 because the binary value “10000000” represents -128 decimal). A static unsigned byte variable called “count” would be declared by the following line:

```
static unsigned char count;        // Create “count” with range of 0 to 255
```

Initialize Variables

Many times you may want to create a variable and start it off at a specific value. This is called initializing a variable. This can be done when you declare the variable. You just make the variable equal to the value you want it to start with. For example let's say you want to declare an 8 bit unsigned variable but want to start it out with the value of 100.

```
unsigned char  value = 100;     // Create variable "value" and initialize it to
                                // 100. The range is 0 to 255
```

You don't have to initialize it when it's declared it's just easier to put it all in one line. You can declare it and then initialize it later with a simple statement as shown below.

```
unsigned char value;            // Create variable "value" range of 0 to 255

value  = 0x64;                  // Initialize "value" to 64 hex or 100 decimal
```

I threw in a curve at you to jump into the next topic. I initialized the variable with a hexadecimal value. As with most compilers you can use different numbering systems. Sometimes hexadecimal is easier to use than decimal. When driving I/O ports I find binary to be useful. In the C compiler you just have to know the format and this is shown in Table 2.

Radix	Format	Example
Binary	0bxxxxxxxx or 0Bxxxxxxxx	0b10011010
Octal	0number	0763
Decimal	number	129
Hexadecimal	0xnumber or 0Xnumber	0x2F

Table 2: Radix Format

The format of the number, known as the Radix, is very common amongst multiple different C compilers. If you've done any programming you will probably already recognize these formats. I find slight differences in the way numbers are formatted when switching from Assembly or BASIC but it's not a major difference. It can be a source of error though when you are just getting started. Some compilers do not recognize binary Radix. When writing for a PC application binary is probably not needed but in the embedded world it's needed. I have found some C compilers require you to use hexadecimal instead of binary to easily declare a port bit pattern.

Constants

Defining a constant value in C is really as easy as defining a variable. In fact there are two ways to do it in C. One is a preprocessor directive using the #define and the other in a format similar to declaring a variable. The differences are minor and I find the #define is a more common method of constant declaration.

#Define

As mentioned earlier in the book, the #define statement is a preprocessor directive that gets acted on before the compiler starts processing your code. Because of this it doesn't need the semi-colon after the statement to indicate end of line. Each #define is a line of its own. For example if we wanted to create two constants called "on" and "off" and associate them with the values of "1" and "0" respectively then two #define statements would handle that.

```
#define ON  1       // The label "ON" represents the value "1"
#define OFF  0      // The label "OFF" represents the value "0"
```

Another useful constant is to declare a port name with a more meaningful label like below.

```
#define Pushbutton  RA3  // The label Pushbutton represents PORTA pin 3
```

These declarations are really just text replacements that make it easier to understand the code. When the compiler sees the label "on" it will replace it with the value "1" before compiling the code. In the statement below the constant is used to make it easier to understand.

```
while ( Pushbutton = ON)        // same as while (RA3 = 1)
        {
        // code inserted here for when PORTA pin 3 is high
        } //end while
```

You can see how constants can make the code easier to understand although some may find RA3 = 1 just as easy. There will be times when declaring constants can really be helpful. RA3 itself is really a constant pre-defined in the PICC-LITE™ include files. I'll cover more of this in a little bit but this declaration of calling PORTA pin 3 RA3 is not standard across all C compilers. Each compiler has its own format so you need to check this in your compiler manual.

Another way to use the #define is in a calculation. In the example below several constants are created and then those constants are used to create a new constant.

```
#define XTFREQ 4                      //Define crystal frequency
#define PLLMODE 2                     //Define phase lock loop mode
#define FCY  XTFREQ*PLLMODE  //Define internal clock frequency
#define BaudRate  9600                //Define BaudRate
#define BRGVAL ((FCY/BaudRate)/16) – 1 //Define UART register value
```

You will notice that this list of constants have dependency on each other. If a program had these constants declared and the hardware used a 4 Mhz oscillator then the register inside the micro titled BRGVAL (which is a constant predefined by the compiler) would have a value based on that. If you wanted to use this same program but with an 8 Mhz oscillator then all you would need to do is change the first #define statement from "4" to "8". This is an example of where the #define statements can make a program much easier to read and modify. Also notice that the constants are capitalized. It's somewhat standard to capitalize the first letter or all the characters of a constant's label to indicate to the person reading the code that a constant value is being used.

Most C compilers are case sensitive so a variable "ON" is not the same as a variable "on". This is why I capitalize constants to make it easier to recognize a constant from a variable and not have to worry about using the same name. Capitalizing constants is a C language standard practice. If you declared a variable "on" and then later forgot and used a constant "on", all your variables named "on" would be replaced by a constant value and probably result in an error that would be difficult to understand. It may even compile without errors and just not function correctly. Many of the BASIC compilers are not case sensitive so if you are moving from BASIC to C then this should be something to watch out for.

CONST

Another option for creating a constant uses the same format used for creating a variable. This is compiler dependent so verify this with the compiler you decide to use but the PICC-Lite C compiler used in this book allows this form of constant creation. It's really treated like another storage class by putting the keyword "const" in front of the constant size. For example to create a byte sized constant for a counter value it would look like this below.

```
const char Counter = 100;        // Constant for starting the counter at 100
```

I get a little confused by this since the compiler will replace the word "Counter" with the fixed value 100 prior to compiling so this declaration looks too similar to declaring a variable if you don't notice the keyword "const" or you forget to capitalize the label. If the statement below is written then you will get an error at compile time.

```
Counter = 150;      // Error! Counter is a constant
```

You also have to include the semi-colon with this constant declaration which may make things more consistent but my thought is statements without semi-colons are clearly pre-processor directives to the compiler while those without it are for compiling. You can make your own choice but in the projects you see later in this book, I use the #define statement for constants.

Just to further clarify the differences between the two methods, I took a small section from one of the project files and show how I would declare some binary constants using either method.

#Define Method:

```
#define DS1_on 0b00000001          //Create Constant to light DS1 LED
#define DS2_on 0b00000010          //Create Constant to light DS2 LED
```

```
#define DS3_on 0b00000100          //Create Constant to light DS3 LED
#define DS4_on 0b00001000          //Create Constant to light DS4 LED
```

"const" Keyword Method:

```
const char DS1_on = 0b00000001;    //Create Constant to light DS1 LED
const char DS2_on = 0b00000010;    //Create Constant to light DS2 LED
const char DS3_on = 0b00000100;    //Create Constant to light DS3 LED
const char DS4_on = 0b00001000;    //Create Constant to light DS4 LED
```

Internal Registers and Constants

Inside every microcontroller are specific memory locations dedicated to controlling the various features such as timers and analog to digital ports. There are also registers dedicated to the I/O ports for setting the direction of input or output or setting or clearing the pin. These locations are not constant in that they can change their value during run time but they are constant in that they have a very specific location in the internal memory. As I pointed out in the first chapter, every program has to include a .h file specific to that compiler. With PICC-Lite that header file is the "pic.h" file included at the top of every PICC-Lite program. Within this file it calls in another .h file that will be dedicated to the microcontroller you have selected in the development screen when you setup your first project. I'll cover more of that in the hardware setup section but for now just know that the PICC-Lite compiler automatically pulls in that microcontroller's register definitions through the .h file the "pic.h" file calls.

If you were to look at the .h header file called in for one of the microcontrollers such as "pic16f685.h" that is included with the PICC-Lite compiler you will see lots of declarations created for all the special function registers along with the port labels. These are done with a similar method to how we create a constant or variable. Below is a small sampling from the "pic16f685.h" file.

```
// Special function register definitions

static volatile     unsigned char       TMR0        @ 0x001;
```

```
static volatile    unsigned char    PCL       @ 0x002;
static volatile    unsigned char    STATUS    @ 0x003;
static             unsigned char    FSR       @ 0x004;
static volatile    unsigned char    PORTA     @ 0x005;
static volatile    unsigned char    PORTB     @ 0x006;
```

The section above shows a small sampling of the special function register setup. If you ever need to know what keyword to use to access these registers, you just have to look in the microcontroller's header file. If I needed to preload the TMR0 register with a value in my program I would write the line below and the compiler would recognize the label "TMR0" as the Timer 0 register at location 0x001 of program memory because it was already defined in the "pic16f685.h" file that was automatically pulled in by the "pic.h" file at compile time.

```
TMR0 = 0xF0;     // Preload Timer 0 with F0 hex
```

The definitions for all the I/O pins are also declared in the "pic16f685.h" file. Knowing these labels is important to writing software without errors. Each C compiler will probably have a different name for these pins. PORTA definitions are shown below.

```
/* Definitions for PORTA register */
static volatile    bit    RA0    @ ((unsigned)&PORTA*8)+0;
static volatile    bit    RA1    @ ((unsigned)&PORTA*8)+1;
static volatile    bit    RA2    @ ((unsigned)&PORTA*8)+2;
static volatile    bit    RA3    @ ((unsigned)&PORTA*8)+3;
static volatile    bit    RA4    @ ((unsigned)&PORTA*8)+4;
static volatile    bit    RA5    @ ((unsigned)&PORTA*8)+5;
```

Within the "pic16f685.h" file is also a batch of constants. These are the configuration bits required for actually programming the microcontroller. When the chip is loaded with the binary .hex file that the C compiler creates, the programming hardware also has to set some configuration bits or sometimes called fuses within the microcontroller. These bits or fuses setup the microcontroller to run with an internal oscillator or an external oscillator. These fuses also enable or disable the watchdog timer. These fuses also enable or disable the power up timer and the brownout reset timer available on some microcontrollers. The .h file for the microcontroller you are using will have these defined so this is definitely something you will need to look at if you want put that information within the binary file sent to the hardware programmer prior to loading the microcontroller with your program. This is the best way to do it since the setup and software program stay together in one file. The options for the oscillator setup of the "pic16f685.h" file are below.

```
// Configuration Mask Definitions
#define CONFIG_ADDR 0x2007
// Oscillator
#define EXTCLK      0x3FFF   // External RC Clockout
#define EXTIO       0x3FFE   // External RC No Clock
#define INTCLK      0x3FFD   // Internal RC Clockout
#define INTIO       0x3FFC   // Internal RC No Clock
#define EC          0x3FFB   // EC
#define HS          0x3FFA   // HS
#define XT          0x3FF9   // XT
#define LP          0x3FF8   // LP
```

Arrays

One final area of declaration I wanted to cover was arrays. These are just simply sequential blocks of RAM space that automatically end with a null character (\0) added to the end. A byte array of 10 elements would take up 11 bytes of space; 10 bytes of data and then the null character. Arrays make it easy to build small lookup tables since you can address each location through the reference number associated. The example below creates an array of 5 elements and preloads it with the five decimal values of 1,2,3,4 and 5.

```
const char Id[] = {1,2,3,4,5};    // Create and initialize constant array
```

The first constant is located at location zero so the statement below would make the variable "x" equal to 1.

```
x = Id[0];    //Make variable "x" equal to value at location 0 of array
```

Notice how I didn't have to specify how many locations of memory I wanted the "id[]" array to take up. The C compiler will automatically make room and then add the null character at the end. You can specify the amount in the brackets if you want but make sure you allow for the null character. An example is below.

```
const char Id[6] = {1,2,3,4,5};
```

Arrays are very useful for storing strings of ASCII characters. You just need to put the characters in quotes to represent string characters. To store the ASCII values for 1,2,3,4 and 5 in an array the statement below will do that. A null character is added at the end.

```
const char Id[] = {"12345"};  //Store ASCII values in array
const char Letter[] = {"Hello"}; //Store ASCII string in array
```

Chapter 4 – Statements

I've already presented some of the C language statements in previous chapters without explaining them. Now I will try to explain the predefined statements for the C language. If you are coming from an assembly or BASIC language background, you may think of these as commands but in C they are called statements. The first one I'll to cover is the more popular *while* statement.

While Statement

The *while* statement is used quite often in a C program to form the main loop of code. It takes on the format seen below.

```
while (expression)
        {
        // Code to run when expression is true
        }
```

The expression can be any simple variable or constant or formula that is tested to see if it results in a zero (false) or not zero (true). If the expression was a simple test to see if one equals one then it will always result in a true expression because one will always equal one. This means all the code within the curly brackets would run over and over again creating a continuous loop. That *while* statement would look like the section below.

```
while ( 1==1 )
        {
        PORTC = 0  // Make all pins of PORTC low or off
        }
```

This is a common method of creating the main loop. At the top of main loop, place a "while (1==1)" and then everything within the curly brackets will run continuously. If you are coming from the BASIC compiler world this would be the same as creating a label "main" and then at the bottom of the main loop place a "goto main" command to make the loop continue forever.

Every embedded C program has to have some kind of continuous loop because it doesn't have an operating system creating a loop like a PC would have.
If instead of "1==1" we used a expression that tests a variable "x" to see if it is equal to one then the program would look like the section below and would not run continuous unless the variable "x" continued to equal one.

```
while ( x ==1 )
        {
        PORTC = 0  // Make all pins of PORTC low or off
        }
```

The variable "x" would have to be declared and then something would have to change it from one to stop the PORTC operations from running. The expression can do things other than test for equality. You can test for x<1 or x>1 or use a variable name that makes more sense like the *while* statement below.

```
#define  switch  RA0
#define OFF  0

while (switch == OFF)
{
PORTC = 0 // Make all pins of PORTC low or off
}
```

You can use #define to make the label "switch" represent the port pin RA0 and the constant "OFF" represent the value zero. In the *while* statement expression the value of switch is tested against the value of OFF. This is tricky if you think about it because we are testing to see if switch is zero. You might think; if it's zero doesn't that mean the *while* statement won't execute the code in the curly brackets? The answer is no because the switch port pin can be high or low based on the position of the switch. If the switch signal is low or digital zero then it will equal the constant "OFF" value of zero so the expression will be true (0 == 0) and the *while* statement will run the code between the curly brackets. If switch was a high signal or 1 then the expression (1==0) will not be true so the code in the curly brackets will not execute.

Testing an expression for true or false can be tricky so I recommend you make the expressions in your *while* statements as clear as possible so you don't execute code at the wrong time.

Notice how the expression is always tested first and then the decision is made to run or not to run the code between the curly brackets. This is because the *while* statement is placed above the code. You can also place the *while* statement at the end.

Take the example below where PORTC is set to zero and then the *while* statement is encountered.

```
main ()
{
PORTC = 0  //Make all pins of PORTC low or off
}
while (1==1); // Stay here forever
```

The PORTC register will be set to zero and then the *while* statement is encountered. Because there is nothing below the *while* statement and the expression is true (1==1 is true) the program will stay at that one command line forever. Notice how the curly brackets are gone and replaced by a semi-colon. This is because this is really an abbreviation of the *while* statement. It's the same as the expanded version below with the curly brackets removed.

```
while (1==1)
{
;               // null statement (explained later in the chapter)
}
```

Do-While Statement

The *do-while* statement is very similar to the *while* statement with one major difference. The *do-while* statement will perform one loop of code within the curly brackets before testing to see if the *while* expression is true or false. If the test proves true, then it will loop back up to the *do* portion and then test again. If the test proves false, then the statement is finished and the program moves on to the next set of statements or functions.

You would use the *do-while* when you want to complete a task at least once before testing an expression for a true or false condition. For example you may want to take a sensor reading the first time to establish a baseline value and then test to see if a timer has expired before taking the next reading.

The *do-while* statement format is shown below.

```
do
        {
        //Code to run entered here
        }

while(expression);
```

The format is very similar to the *while* statement format except the while is moved to the end and the *do* is at the top. This is similar to any *do-while* statement you might see in BASIC language.

For Statement

The *for* statement is often referred to as the *for – loop* statement. It is very similar to the BASIC For – Next command. The *for – loop* is just another way to execute a block of code over and over a specific number of times. The difference between this command and the BASIC language version is more about the structure than the way it is evaluated. The C language *for – loop* will continue as long as a defined expression is true where the BASIC language For – Next will continue until the variable reaches a specific value. Let me show the format first before I explain that better for those coming from the embedded BASIC arena.

```
for ( variable = initial value; expression; variable = variable + value)
        {
        // Code to execute while expression is true
        }
```

That's the format, now let me show you a real example. The example below initializes the variable "x" to the value one. Then it tests if the value is less than nine. If the value is less than nine the result is true and the code between the curly brackets is performed. In this case the binary value of "x" is placed in the PortC register of the microcontroller. In this example it is assumed that LEDs attached to the PortC pins will light up when a one is at their location.

```
for(x=1; x<9; x=x*2)
      {
      PORTC = x;                // Turn on next LED
      }      // End For
```

After the last command line between the curly brackets is completed, the *for-loop* will change the value of the "x" variable based on the third element of the *for* statement. In this case the value of "x" is multiplied by two and then placed back into the same variable location "x". In other words, the value of "x" doubles after every loop. Before executing the code between the curly brackets though, the expression is tested again to see if "x" is still less than nine. If it is, the expression is true so the commands are executed again. This time different LEDs are lit. This continues until x is doubled to equal a value larger than nine. When x is larger than nine the expression will be false and the program execution will move to the statements after the lower curly bracket (which is not shown in the example above).

For Embedded BASIC programmers, this can add a bit of confusion. In the typical BASIC language command setup the initialization and the expression are combined in the BASIC For-Next command line. The incrementing of the variable is done at the Next command line. For example, in the PICBASIC PRO For-Next command below the loop will continue until the value is greater than nine.

```
For x = 1 to 9
' Commands entered here
Next x
```

In C language this would simply be written as follows:

```
for (x = 1; x<=9; x = x+1)
```

```
{
//Commands entered here
}
```

Hopefully you can see that converting to C from another language such as BASIC just requires a bit of patience until you learn the formatting.

Shortcuts are another area of confusion. I often see the x = x+1 replaced by x++ in the *for* command lines. I still prefer to spell it out so anybody with beginning programming skills can understand what is happening.

If Statement

The *if* statement also has a similar command in the BASIC language where it is typically called an *if-then* command. The format for the C language version is below.

```
if (expression)
    {
    // Code executed if expression is true
    }
```

A simple example where the *if* statement might be used is to test a port to see if it is low. This is a common practice often used to test the state of a momentary switch. It is common to have the port pulled high by a pull-up resistor to voltage and then when the normally open switch is pressed, the port voltage drops to zero and a low logic level can be sensed by the port (Reference Figure 1).

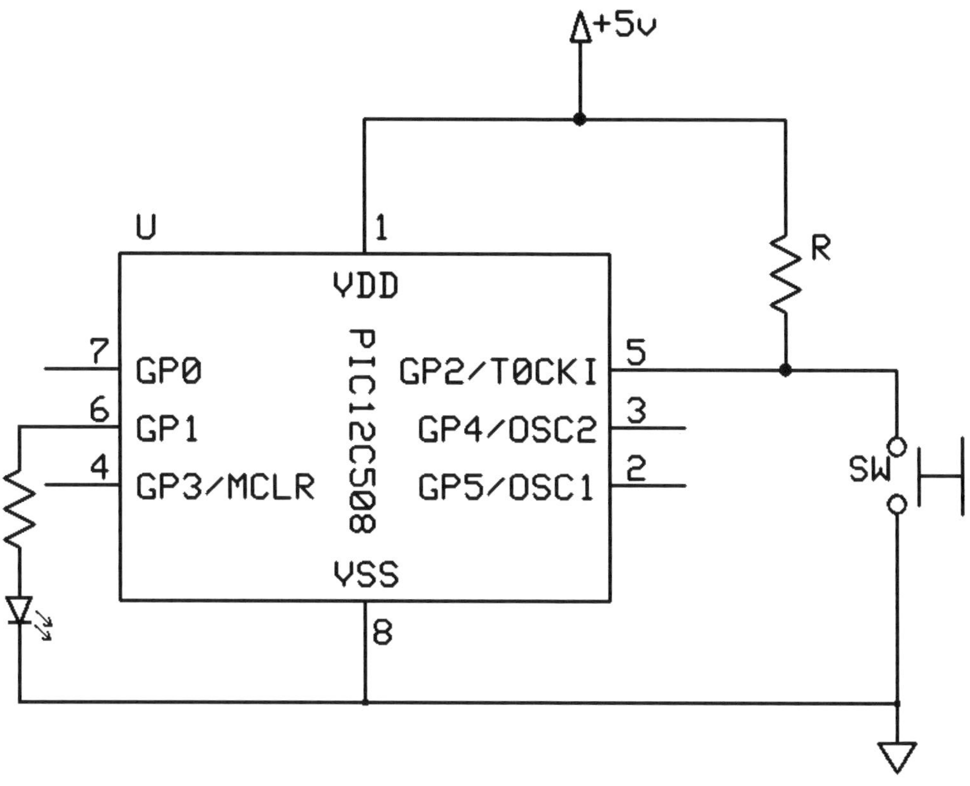

Figure 1: Switch Schematic

The C code to read the switch might look like the example below. The *if* statement tests the level of the GP2 pin to see if it is low or a logical zero. If that statement is true or the switch is pressed, then the output port GP1 is set to a logical one or high level that lights the LED.

```
if (GP2 == 0)            // Test RA3 port
      {
      GP1 = 1;           // If SW1 pressed, turn on LED
      }
```

If you look at it, this statement is not much different than the *while* statement. The *if* statement has more options though.

If-Else

In many applications you will have at least two modes to respond to. In my previous *if-then* example of reading the switch, I didn't show what the microcontroller would do if the switch is tested and is found to be a high or logical one level indicating the switch was not pressed. A simple *else* statement can be added to handle that as seen in the example below.

```
if (GP2 == 0)               // Test RA3 port
      {
      GP1 = 1;              // If SW1 pressed, turn on LED
      }
else
      {
      GP2 = 0;              // If SW1 not pressed, turn off LED
      }
```

If the expression "`GP2 == 0`" results in a false condition (which happens when the port is high or at a logical one) when the switch is not pressed, then the code after the *else* statement is executed.

Else-If

Another variation is to add a second test to the same section of code. For example, what if you needed to make several different measurements to determine what your action should be. You could add another *if* statement but still based in the same block of code. For example, what if you wanted to test a variable to see if it is below 100, above 100 or equal to 100? The example below uses the *else-if* statement to add that third level of test.

```
if (x < 100)                // Test variable x for less than 100
      {
      GP1 = 1;              // Turn on LED1
      GP0 = 0;              // Turn off LED2
      }
else if (x > 100)           // Test variable x for greater than 100
      {
      GP1 = 0;              // Turn off LED1
      GP0 = 1;              // Turn on LED2
      }
else                        // Assume x = 100
      {
      GP1 = 1;              // Turn on LED1
      GP0 = 1;              // Turn on LED2
      }
```

The program will light two LEDs based on the value of variable "x". If the value of "x" is less than 100 then LED1 is lit. If "x" is greater than 100 then LED2 is lit. If "x" is equal to 100 then both LEDs are lit. This also leads us into the next topic, the *switch-case* statement which performs this type of function with multiple choices

Switch-Case

The *switch-case* statement is an option for situations where you might need several *if-then* statements. The format for the *switch-case* statement is shown below.

```
switch (variable)
{
case constant0:
        ; //code for action if variable = constant0
        break;
case constant1:
        ; //code for action if variable = constant1
        break;
case constant2:
        ; //code for action if variable = constant2
        break;
.
.
.
default:
        ; //code for action if variable never equals a constant value

}
```

The statement starts with the *switch* statement followed by a variable in parenthesis. The value of the variable is then compared against the list of constants that follow each *case* statement. If the variable matches one of the constants, then the code that follows that *case* statement is executed and the program exits the *switch-case* statement. If the value of the variable doesn't match any of the constants then the code after the *default* statement is executed. This could be an error message or just a reset of the variables value back to zero.

You can think of the *switch-case* statement as a way to create a look-up table to select a different set of messages to display on an LCD or how to light LEDs in a different set of patterns. In fact I'll use this in chapter 14 to light the LEDs on the PICkit 2 development board I use in this book. A snippet of the code is shown below to select which LED to light based on the value of the variable "state_led".

```
switch (state_led)
             {
                  case 1:              // STATE0: turn only the D0 LED on
                        RC0 = 1;
                        RC1 = 0;
                        RC2 = 0;
                        RC3 = 0;
                        break;
                  case 2:              // STATE1: turn only the D1 LED on
                        RC0 = 0;
                        RC1 = 1;
                        RC2 = 0;
                        RC3 = 0;
                        break;
                  case 3:              // STATE2: turn only the D2 LED on
                        RC0 = 0;
                        RC1 = 0;
                        RC2 = 1;
                        RC3 = 0;
                        break;
                  case 4:              // STATE3: turn only the D3 LED on
                        RC0 = 0;
                        RC1 = 0;
                        RC2 = 0;
                        RC3 = 1;
                        break;
                  default:
                        state_led = 0; // If state_led > 3 reset
                                       //  switch_led to zero
                        RC0 = 0;       // All LEDs off
                        RC1 = 0;
                        RC2 = 0;
                        RC3 = 0;
                        break;
                  } //end switch
```

Break Statement

The *break* statement is used to exit a loop of code or section of code. It can be used anywhere and will jump the program control to the next command line after the set of curly brackets containing the *break* command. As you have just seen, it is used to end the set of commands in each *switch-case* option. It can be used to exit a *for-loop* when a certain value is detected.

```
for (x = 1; x<=9; x = x+2)
        {
        PORTC = x                       // Set PORTC
        if (x ==7)                      // Exit if x ever equals 7
                break;
        }
```

This can be useful if you want to exit on an error condition. In the example above, x should never equal seven but if it does the loop will be exited.

Continue

The *continue* statement is the opposite of the *break* statement. When the *continue* statement is encountered it doesn't leave the curly brackets, it just skips all instructions after the *continue* statement and returns program control to the top of the same curly brackets. It's a way to skip the rest of the statements within the curly brackets under certain conditions and not leave the loop.

```
for (x = 1; x<=9; x = x+1)
        {
        PORTC = x                       // Set PORTC
        if (x ==7)                      // Bypass pause if x ever equals 7
                continue;
        pause (500)                     // Call function pause
        }
```

The example above calls the function *pause* but bypassed it if x ever equals seven.

; - Null Statement

The last statement I wanted to mention is the *null* statement. If you program in assembly you can think of this as the NOP command but without losing any instruction cycles. The *null* statement just marks a point in the program where the compiler needed another statement for syntax purposes. The *null* statement isn't even a statement really; it's just a semi-colon character.

As I mentioned early on in this book, semi-colons are required to end a command line and it's one of the things a beginner will often forget. You will get compiler errors when you leave them off and the error message won't state clearly that you simply left off a semi-colon.

If you think of the semi-colon as a *null* statement, then adding the semi-colon after each command line starts to make more sense. It's like the old fashioned way of communicating by sending a telegram. After every sentence the reader would add the word "stop".

I'll be home at 7. Stop
I will bring a pizza. Stop
Have a cold one ready for me. Stop

The *null* character is that stop statement for your C language command lines. The only time you don't need to add a semi-colon to a command line is when the statement used in the command line ends with a "}" curly bracket. Here is trick though for those of us with poor memory. If you add a semi-colon to the end of every command line, even those with curly brackets then the compiler will either accept the semi-colon when needed or just treat the extra semi-colon as a *null* character when it's not needed. This is a safe way to eliminate those pesky semi-colon created errors.

Below I reproduced the simple *if-then* statement example used earlier but this time I added a semi-colon to the end of it. Since *if-then* ends with a curly bracket, it doesn't need a semi-colon at the end. If you play it safe and add one, the compiler will just treat it as a *null* statement.

```
if (GP2 == 0)               // Test RA3 port
      {
      GP1 = 1;              // If SW1 pressed, turn on LED
      }
else
      {
      GP2 = 0;              // If SW1 not pressed, turn off LED
      };           // Extra semi-colon is treated as NULL statement
```

This covers the major statements you will come across in most C programs and is a big enough list to get you started with the C language. Everything I've covered will be used in this book's project section. If you understand these statements you can understand the guts of most C programs. The shortcuts associated with some of these can cause confusion and I purposely didn't cover those. I want to teach you how to write your own programs that are easy to understand.

Chapter 5 – Mathematical and Logical Operations

Solving math equations in a single command line is one advantage a compiler offers when compared to programming in Assembly. C compilers offer a lot of options for solving math equations. Floating point math is very common with C compilers but that won't be covered in this book as I feel that is an advanced topic. Floating point math isn't difficult to understand but how that all works within an 8-bit microcontroller can be very confusing for the beginning embedded developer. The chart below shows some of the most common mathematical operator symbols.

*	Multiplication
/	Division
+	Addition
-	Subtraction
%	Modulus or Division Remainder
=	Equate or set equal to

```
c = a        // c is set equal to the value of a
c = a * b;   // c is the result of a multiplied by b
c = a / b    // c is the result of a divided by b without remainder
c = a % b    // c is the remainder of a divided by b
c = a + b    // c is the result of the value of a added to the value of b
c = a – b    // c is the result of the value of b subtracted from the value of a
```

Shortcuts

Math expression shortcuts are an area of C programming that confused me the most and to this day I cannot understand why there is such a focus on all the shortcuts. For example, what is the difference between equation 1 and equation 2?

Equation 1

b = a++ ;

Equation 2

```
b = ++a;
```

If you are new to C programming you are probably looking at these statements and scratching your head thinking those don't make any sense to me? These are examples of mathematical shortcuts that you will find in 90% of all C programs. I for one can live without them but if you want to understand C language you have to understand them.

The clue you may be missing is that the expression a++ is the same as a = a+1. The second form took 5 keystrokes while the first one took three. The argument I've always heard for C shortcuts is less typing required. My argument is you will be required to do more typing in the comment section if you want anybody including beginners to understand your code. This is important to me but some don't see the advantage. I've also seen many poorly commented code samples. Let me write Equation 1 again and then expand it out the way I prefer to write it.

```
b = a++;      // Is the same as b = a and then a=a+1
```

or

```
b = a;        // No comments necessary
a = a + 1;
```

My method took less keystrokes because of all the comments but a more experienced programmer probably would not have written all those comments I added to the first method so overall the shortcut would be less keystrokes. If I expand out Equation 2 you may begin to see why I prefer the expanded method.

```
b = ++a;      // Is the same as a = a+1 and then b = a
```

or

```
a = a + 1;    // Again, no comments necessary
b = a;
```

The difference between Equation 1 and Equation 2 is the order of operation. I would prefer to just write it out rather than have to remember if the ++ signs preceding or following the variable mean increment before or after performing the math function. Here are a fcw more to confuse you.

```
c = a+b++;  // This is a legitimate expression
```

This is the same as:

```
c = a + b;
b = b + 1;
```

How about this:

```
c = a+  ++b;
```

It means the same as:

```
b = b + 1;
c = a + b;
```

I searched numerous C manuals and books to fully understand these shortcuts and I've found numerous summaries but I still prefer to just spell it out even if it takes more keystrokes.

These shortcuts also exist for decrementing.

```
b = a--;        // Is the same as b = a and  then a = a-1

b = --a;        // Is the same as a = a-1 and then b = a
```

Relational Operators

You have numerous ways to test the relation of two values or variables. The table below shows the most popular.

Operator	Meaning
>	Greater than
>=	Greater than or Equal to
<	Less than
<=	Less than or Equal to
==	Equal to
!=	Not equal to

These operators just test if a statement is true or false represented as a “1” or “0” binary value. You will see these used as qualifiers in statements such as a *while* or *do – while* statements.

The section of code below will continue looping until the value of *a* is equal to zero. When the value of *a* is equal to zero then the statement a != 0 (a is not equal to 0) is considered false because *a* is equal to zero so the binary result of the expression is zero or false. The *while* statement and anything within it’s curly brackets will continue until the expression in the *while* statement is false.

```
unsigned char a = 255

while ( a != 0)
        {
        a = a - 1;
        } // end while
```

Another area of confusion for the beginner is the difference between the two statements below.

```
a = 0;
```

and

```
a == 0;
```

The first one is a mathematical operator that makes the value of the variable *a* change to equal zero. You might use this to initialize a variable to a specific value such as *a = 200.*

The second expression tests to see if the variable *a* is equal to zero. If it is equal to zero then a logical one or true is returned, if not equal to zero then a logical zero or false is returned. The value of *a* doesn't change.

Both of these will compile without errors if you get them switched so you may be staring at your program and find it hard to see why it's not working only to later find out you used the wrong number of equal signs. I know I have used only one equal sign before when I meant to test the variable with two.

Logical Operators
Logical operators are similar to relational operators in that they produce a true or false response. The table below shows the logical operators.

		AND	OR	NOT
		&&	||	!x
a	b	a && b	a || b	!a
0	0	0	0	1
0	1	0	1	1
1	0	0	1	0
1	1	1	1	0

These are typically used the same way as relational operators, to test if an expression is true or false before leaving a *while* or *do-while* loop.

Bitwise Operators

Bitwise operators actually change the value of a variable. The options are in the table below.

&	AND
|	OR
^	XOR
~	1's Compliment
>>	Right Shift
<<	Left Shift

These can be used to look at specific bits of a variable. For example lets assume that the variable "a" equals binary 0b01011010. If we AND it with 0b11110000 we can see the upper nibble. If we AND it with 0b00001111 we can see the lower nibble.

```
a = 0b01011010;
b = 0b00001111;

c = a & b;          // c = 0b00001010 or 0x0A

b = 0b11110000;

c = a & b;          // c = 0b01010000 or 0x50
```

This is a handy method of breaking out just portions of a microcontroller port. If you read in the whole port of eight pins but only want to look for action on a few of the pins, this is a great way to specify which to read and which to ignore. You probably won't do this very often on PC programming but you will use it often in embedded C programming.

Bit shifting is also handy. The statements below will shift the bits around.

```
a = 0b00001111;
b = a << 4;         // b equals 0b11110000 as bits in "a" were shifted 4
                    // spaces to the left
```

You can also see shortcuts with logical operators. This is another area I suggest you spell out the whole equation to avoid beginner confusion. Remember, you will have to go back and understand what you were thinking at some point after the code is written and programmed into a chip. This is especially true if you want to modify what the program did in the first place.

Bitwise shortcut:

```
b = &a;             // b = a & b
```

Precedence

Just like the math you learned in school, all the mathematical expressions have an order of precedence. If part of the equation is in parenthesis then that will be performed first. The rest follow the normal structure. Multiplication precedes division and then addition followed by subtraction. Logical expressions get inserted in there as well. It's best to consult the C compiler manual for the exact order.

In fact I suggest you pull out the C compiler manual for the PICC-Lite when you have time and read about the mathematical operations. Most C compilers offer many more options than I've listed here. I just wanted to cover the basics.

Chapter 6 – Functions

Functions are what make C programming the preferred language based on my experience. If you have written programs in another language you may think of functions as subroutines. There are very few predefined commands, or statements as they are called in the C language, such as *if-then*, *do-while* and a few others that we covered in Chapter 4. Everything else you put in your C program will be a custom written function. This is also where the portability of the C language gets some of its popularity. If I know someone who has written a C language function to control an LCD module and if they will let me have a copy of that function, I can paste it in my program (or link it in which I will discuss a little bit later) and have instant LCD control.

So the theory goes that if a lot of people choose to program in the C language then there should be a lot of pre-written functions that I can use to make creating a new program easier. The problem comes in when some people don't want to share their functions. After all, they took the time to write them, test them and prove them out. I can't really blame them. So maybe you can buy functions for a low cost? This doesn't seem to be a great opportunity either as the person paying money for it will require documentation on how to use the function and how it works and I've found most people really aren't interested in writing all the details about their code for a few dollars of profit.

Most C compilers will include several pre-written functions and group them together in a file known as a library with some instructions on how to use each one. Some C compilers come with open source functions that you can easily modify and others come with protected functions to keep them standard. If you've used one of the BASIC compilers you will have lots of pre-written commands to choose from. These commands are just pre-written functions that are typically closed to modification or at least not easy to modify without detailed knowledge. As you write more programs you will find your collection of custom functions will grow.

There are specific elements to understand before using functions and they include; function prototype, function arguments and function return.

Function Prototype
The function prototype is a one line description of the function that is declared at the top of the main program just to let the C compiler know it is coming. This can actually be skipped if the full function is placed before the main loop of code. Some people prefer to have all their functions written in blocks at the top of the program and then the last loop of code is the main loop. This can become a very long program but for simple applications this can work well. In this situation, a prototype does not need to be included.

Another option is to put all the function prototypes together in a .h header file that can be included at the top of the program. This way all functions are declared at once and you have a single file to modify if you want to add, delete or modify any function prototypes. The format for a function prototype is as follows:

type function_name (type variable1, …);

Let's assume we need a function prototype declared for a time delay function called "msecbase". The example below shows a simple function prototype declared for the function named "msecbase". This function will actually be used in the Chapter 10 project. Notice that the prototype ends with a semi-colon.

```
void msecbase( void );          //Establish millisecond base function
```

The "void" at the front and within the parenthesis are just place holders. The first void represents the returned value type which could be an int or char variable but in this case we don't need to return anything so we use void. We also don't need to pass any values to the function for it to use in its calculations so the void in the parenthesis replaces any variable declarations.

This function prototype declared at the top of the program must match the first line of the actual function. The full function is shown below. The function starts off with a full description of what it does within a comment header.

```
/*******************************************************
* msecbase - 1 msec pause routine                      *
* The Internal oscillator is set to 4 Mhz and the      *
* internal instruction clock is 1/4 of the oscillator.  *
* This makes the internal instruction clock 1 Mhz or    *
* 1 usec per clock pulse.                               *
* Using the 1:4 prescaler on the clock input to Timer0  *
* slows the Timer0 count increment to 1 count/4 usec.   *
* Therefore 250 counts of the Timer0 would make a one   *
* millisecond delay (250 * 4 usec). But there are other *
* instructions in the delay loop so using the MPLAB     *
* stopwatch we find that we need Timer0 to overflow at  *
* 243 clock ticks. Preset Timer0 to 13 (0D hex) to make *
* Timer0 overflow at 243 clock ticks (256-13 = 243).    *
* This results in a 1.001 millisecond delay which is    *
* close enough.                                         *
*******************************************************/

void msecbase(void)
      {
      OPTION = 0b00000001;      //Set prescaler to TMRO 1:4
      TMR0 = 0xD;               //Preset TMRO to overflow on 250 counts
      while(!T0IF);             //Stay until TMRO overflow flag equals 1
      T0IF = 0;                 //Clear the TMR0 overflow flag
      }
```

The first code line is the same line of text used in the function prototype except it doesn't need the semi-colon. The code between the curly brackets performs the one millisecond time base

Function Arguments

Sometimes you may want to pass data to the function to use in its calculations. We can do that by creating a variable in the declaration of the function. Instead of "void" between the parentheses of the function declaration an argument is declared. The argument will be a local variable or variables. To demonstrate this I will create a new function called *pause* that will call a second function called *msecbase*. The *msecbase* function will be a routine that delays a fixed amount of time equal to one millisecond. The *pause* function will call that *msecbase* function a specific amount of times to create a larger delay. That specific amount of time will be based on the value the main program sends to the function argument of the *pause* function. First the function prototype gets declared at the top of the program and looks like the line below.

```
void pause( unsigned short msvalue );//Establish pause routine function
```

Notice that an unsigned short (16 bit) argument named *msvalue* is declared. This is the function argument. The first line within the function creates a local variable *x* and then the function uses a *for- loop* to call the *msecbase* function based on the value of *msvalue*. Each time the *for- loop* is incremented it is tested against the functional argument value passed to the variable *msvalue*. If the value of *x* is equal to or less then the value of *msvalue* then the statement is true and another call to the function *msecbase* is performed. As you can see, the value passed to the *pause* function controls how long the delay is. This set of functions can be used to create several different delays within the same program. The *pause* function is below.

```
//**************************************************
//pause - multiple millisecond delay routine
//**************************************************

void pause( unsigned short msvalue )
{
      unsigned short x;

      for (x=0; x<=msvalue; x++)//Loop through a delay equal to usvalue
            {                   // in milliseconds.
            msecbase();         //Jump to millisec delay routine
            }
}
```

You may still be thinking how do we use this to create a delay in our main loop? If the main loop of code wanted to delay 10 milliseconds then it would call the *pause* function and pass it a value of 10 with the following command format.

```
main ()
{
pause (10);
}
```

You can pass multiple values to the function as long as the function has declared multiple arguments. This can be really handy for creating custom math functions. You can pass the numbers in an equation to a function that takes those numbers and performs a custom mathematical calculation and then returns the result. In fact let me discuss that next.

Function Return

A function can do more than just receive data, it can use it and then send back a result. Sending data back to the routine that called the function is called a function return. Let's create a simple function to add two numbers together and send back the result as an int value. The function prototype will be the same as the first line of the function and would look like the function prototype below.

```
char add_two_numbers (char i,char j);
```

The return type is declared as a "char" as seen at the front of the function prototype. If we wanted a 16 bit result we could use an "int" type. The prototype also declares two "char" arguments i and j. The function is given the descriptive name "add_two_numbers". Now we create the actual function with a very simple math equation.

```
char add_two_numbers (char i, char j)
	{
	char w;		// Create local variable for storing result
	w = i + j;	// Add the received values
	return w;	// Send the result back
	}
```

The *return* statement sends back the result of the addition as an eight bit value. To use this function within the main program we would issue the following command sequence within the main loop.

```
char x,y,z;  // Create three eight bit variables
char add_two_numbers (char x, char y); // Function prototype

main()
{

for (x=10; x<=100; x=x+10)   // x ranges from 10 to 100
	{
	for (y=0; y<=10; y=y+1) // y ranges from 0 to 10
		{
			z = add_two_numbers (x,y); // Call the add function
		}//end for y
	}//end for x


}//end main
```

```
//*******************************************************
//add_two_numbers - Add Two Numbers Function
//*******************************************************

char add_two_numbers (char i, char j)
      {
      char w;        // Create local variable for storing result
      w = i + j;     // Add the received values
      return w;      // Send the result back
      }              //End function
```

This is a very simple program but does demonstrate function returns. The program will give the following results the first three times through the main loop.

z = 10 (x = 10, y = 0)
z = 11 (x = 10, y = 1)
z = 12 (x = 10, y = 2)

There is a shortcut I wanted to point out in this code. When I created the variables x, y and z, I did it with one command line: char x,y,z; This was probably not a good idea as I'm trying to teach from the beginners perspective. Since I've already criticized the a++ shortcuts, I'm not being fair. I do have a purpose for this though. I try to show everything in this book at the basic level but even then you will find shortcuts and other ways to create programs and functions that work equally as well. There is no one right way to program and the C language is very forgiving and open to new methods. Hopefully I teach you enough so you can figure these out on your own.

The function could have been simplified as well. I didn't need to create the "w" local variable for the result. I could have just used the command line:

return i + j; // shortcut that also returns the same result as w = i + j;

I prefer to spell it out though so there is no confusion in my code when I look back at it months from now.

As you can see, functions are very useful. I will use them often in the projects. We are almost ready to start the projects. First we need to setup the hardware.

Chapter 7 – Project Hardware Setup

I wanted to use a simple hardware setup with this book and I found it in the Microchip PICkit 2 Starter Kit. I'll step you through the setup in this Chapter. If you are a complete beginner then following these steps should make things a lot easier to get up an running quickly. If on the other hand you've used Microchip Technology Inc.'s MPLAB development environment before then you can probably just skim over this chapter and move on to the project chapters. I wanted to make this extremely easy for the beginner because having the hardware setup correctly makes it so much easier to learn how to program. The hardware setup can add a layer of confusion. I think that is why a lot of books on C tend to focus on teaching using a computer for the hardware rather than actual hardware circuitry but, skipping it in an embedded book is like making a peanut butter and jelly sandwich without the peanut butter so I'm including a detailed hardware explanation.

There are many different Microchip PIC Microcontroller programmers to choose from. I have chosen the PICkit 2 Starter Kit for this book because it's inexpensive, comes with a PIC16F690 microcontroller and a development board with LEDs, a switch and potentiometer already soldered in place. The CD included with the package also includes the MPLAB software and the HI-TECH PICC-Lite C compiler. Therefore this package includes everything you need to do the projects in this book and many of the projects you'll create after you finish this book.

The PICkit 2 programmer connects to the PC through a USB port and the USB port also powers the development board. This means you don't have to add any external power supply for any of the projects but you do have the option to power the board separately. The PICkit 2 will sense that extra power and adjust.

The PICkit 2 Starter Kit is shown in Figure 7-1 and can be purchased at various locations including MicrochipDirect.com and many other online stores. If you perform an internet search on "PICkit 2" and you'll find lots of places to buy this package. You can get a package deal on both this book and PICkit 2 Starter Kit at BeginnerElectronics.com.

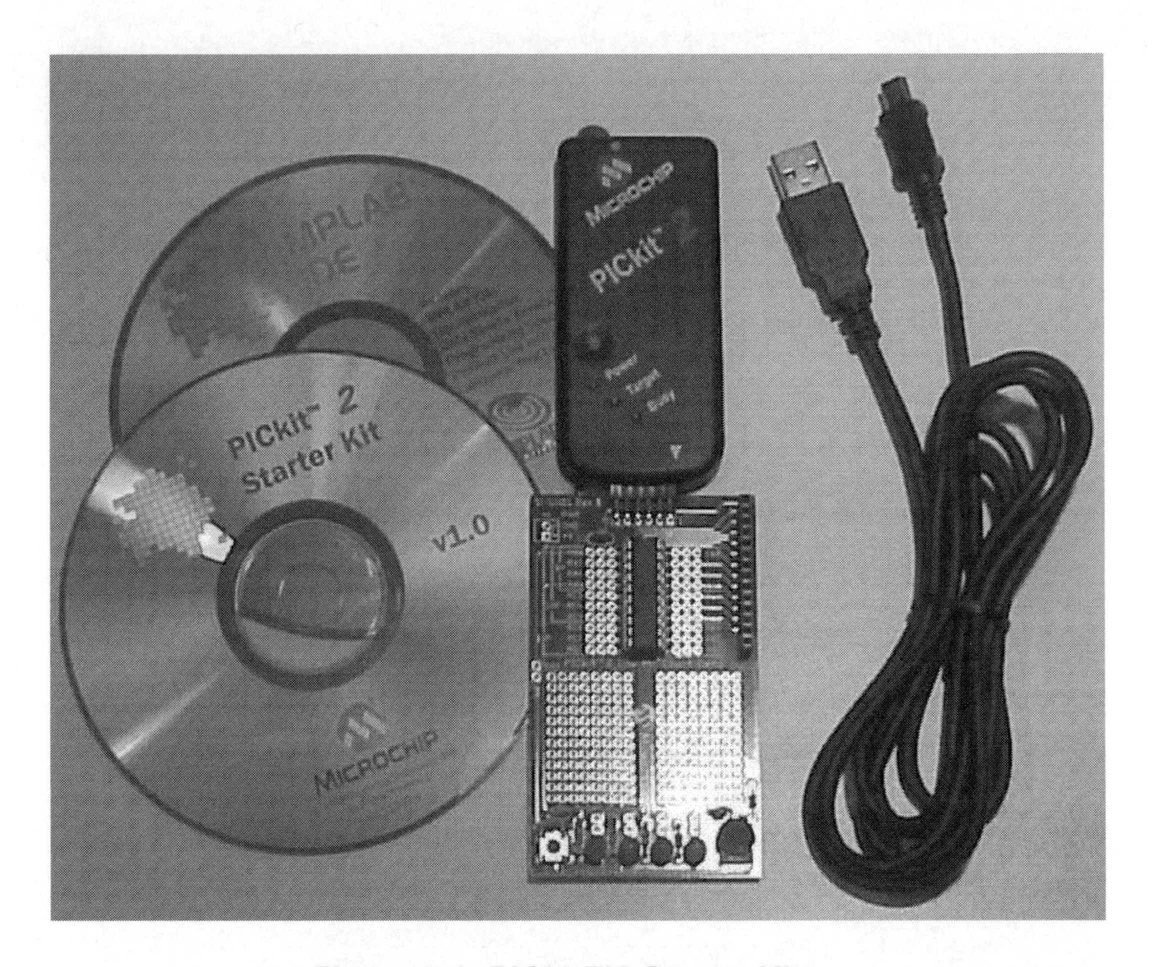

Figure 7-1: PICkit™2 Starter Kit

Project Files
The project software and all supporting files can be downloaded from my website www.elproducts.com/cbookfiles. This will give you everything I reference in this book going forward in a single .zip file. I suggest you unzip

the file into a directory named "Beginner" on your computer hard drive right at the root directory i.e. C:\Beginner. This will keep all path names short and will be easy to find. If you want to put it somewhere else, that is fine but I will reference the beginner directory throughout.

MPLAB Installation

The next step you need to take is to install the latest version of Microchip's MPLAB Integrated Design Environment (IDE) software on your computer. As I write this the latest release is version 8.02. This software installs on a PC but unfortunately doesn't work on a MAC. You can download the latest version of MPLAB for free from the www.microchip.com/MPLAB website. It will download as a .zip file so you will have to unzip it into a directory of your choice on your computer hard drive. When it's unzipped, there should be six files on your hard drive. They are listed below.

instmsia.exe
instmsiw.exe
MP802_install.exe
MPLABcert.bmp
MPLAB Tools v8.02.msi
Data1.cab

The only one you need to run is the MP802_install.exe. The number "802" will be different based on the version of the MPLAB software you downloaded but the format is the same.

When you run the MP802_install.exe file the program will step you through a series of pop-up screens. The first one you should see is in Figure 7-2. This will start the series of screens you can step through as you install the MPLAB software on your computer.

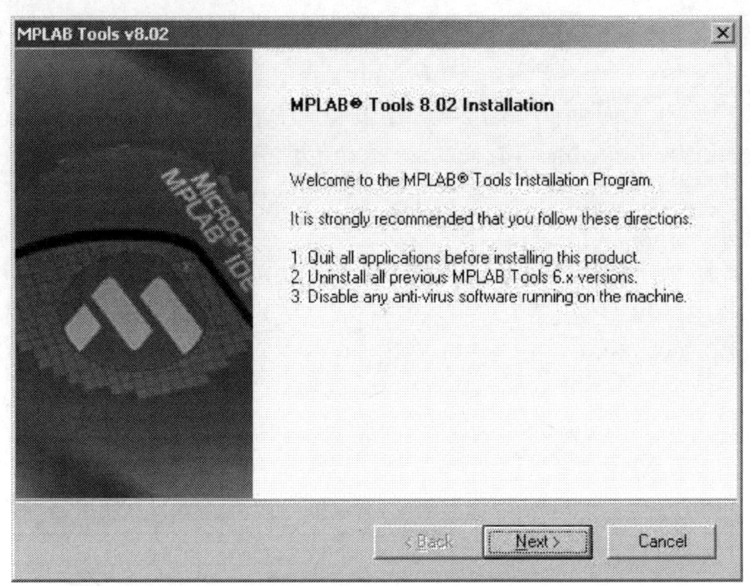

Figure 7-2: MPLAB Installation Screen 1

Figure 7-2 is the welcome screen. Click on the “Next” button to move on but as the screen states, I highly recommend you close all applications you may have running so you don’t get any conflicts. At the end of the installation you will have to restart your computer anyway so why not just shut everything down. Figure 7-3 shows the next screen.

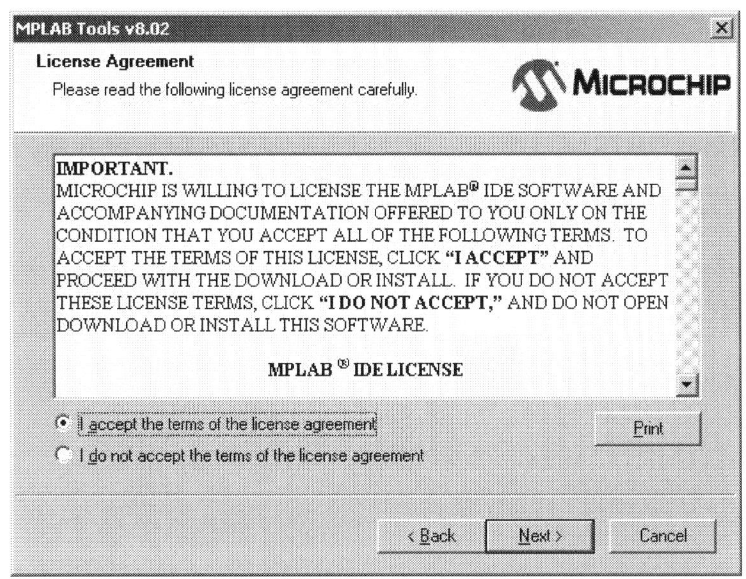

Figure 7-3: MPLAB Installation Screen 2

Read over the license agreement and if you accept it, click in the "I accept" radio button and then the "Next" button to get to the third screen shown in Figure 7-4.

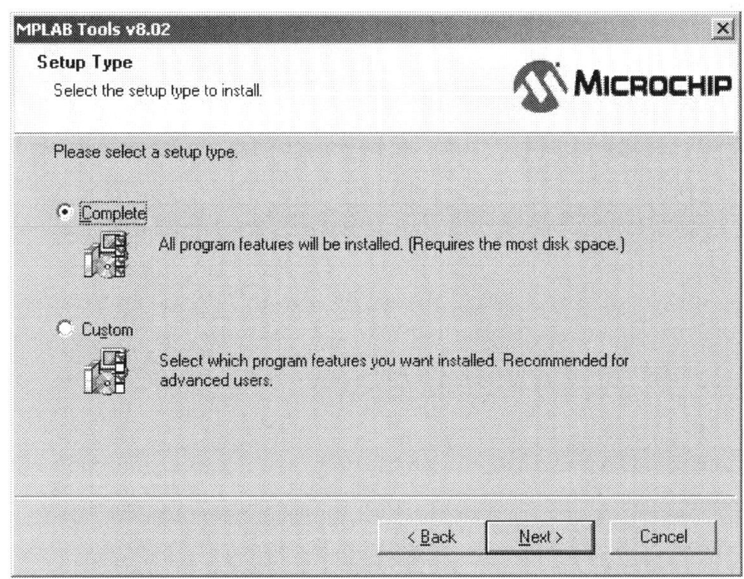

Figure 7-4: MPLAB Installation Screen 3

If you really need to save hard drive space, you may want to select the "Custom" installation but I recommend the "Complete" installation. This will put all the Microchip tools and files on your hard drive so you aren't looking for them later on. Click on "Next" to get to the fourth screen shown in Figure 7-5.

Screen 4 will offer you a path to install MPLAB. It is tempting to put this in another location but I highly recommend you let this install at the default directory. All Microchip documentation that refers to the path will reference this location.

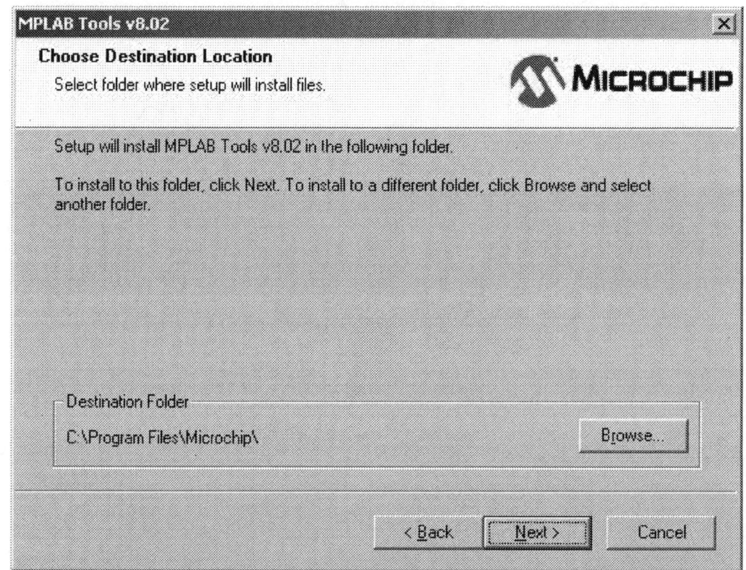

Figure 7-5: MPLAB Installation Screen 4

Click on the "Next" button to get the fifth screen shown in Figure 7-6.

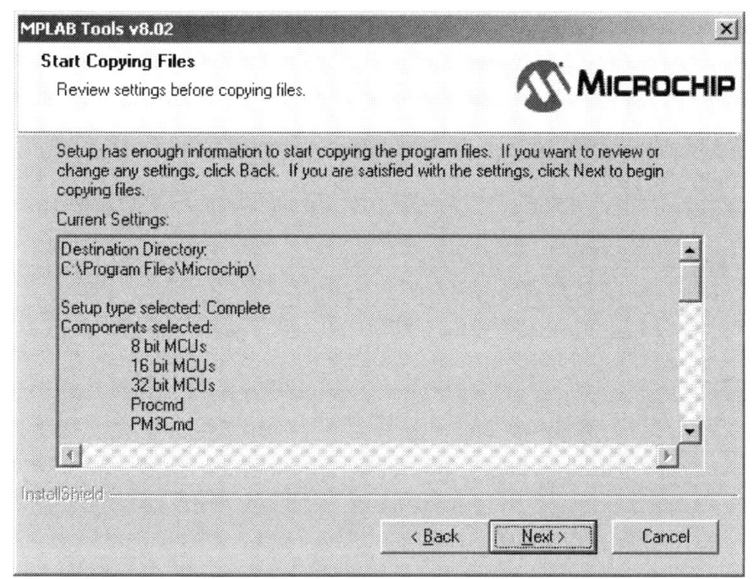

Figure 7-6: MPLAB Installation Screen 5

The fifth screen just lists all the setup information. You can scroll through if you want but when you are ready to move on click on "Next" to get screen six shown in Figure 7-7.

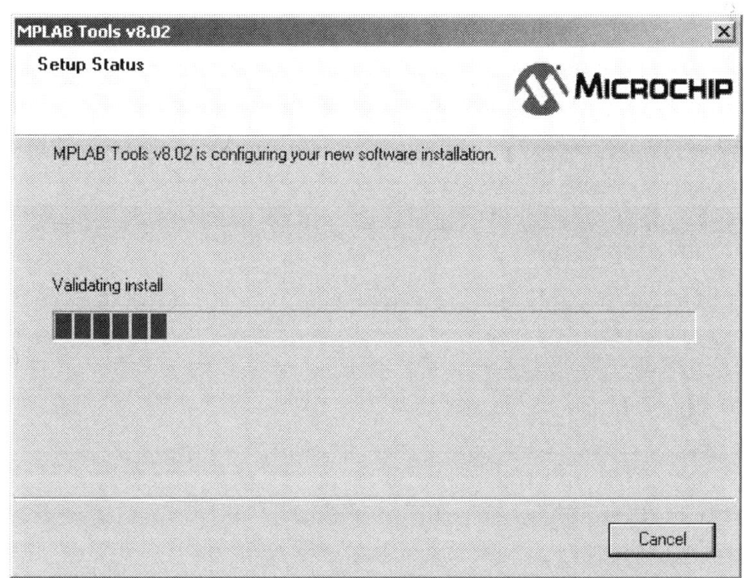

Figure 7-7: MPLAB Installation Screen 6

The installation process will get real slow now as all the files are loaded on your computer. I captured just one of the screens in this process but you will see many very similar screens as different sections are installed. This will continue for a little while so grab a coffee and relax.

HI_TECH PICC-Lite Installation

When all the files have been loaded, MPLAB installation will not end. Instead it will take a twist and offer to automatically install the HI-TECH© PICC-Lite compiler. This is the same C compiler used throughout this book so click on the "OK" button to let it happen.

Figure 7-8: MPLAB PICC-Lite Installation Screen

The installation screens will now switch over from the MPLAB installation to the PICC-Lite installation even though MPLAB installation is not done. Click on the "OK" button to begin the PICC-Lite installation. The image in Figure 7-9 will appear.

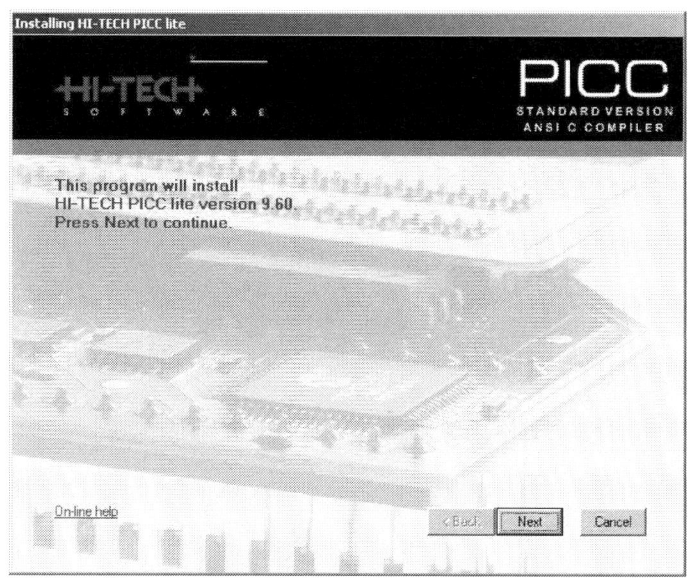

Figure 7-9: HI-TECH PICC-Lite Installation Screen 1

The revision level of the PICC-Lite compiler is shown in this window and it should be the latest level of PICC-Lite if you downloaded the latest version of MPLAB. Click on the "Next" button and the screen in Figure 7-10 should appear.

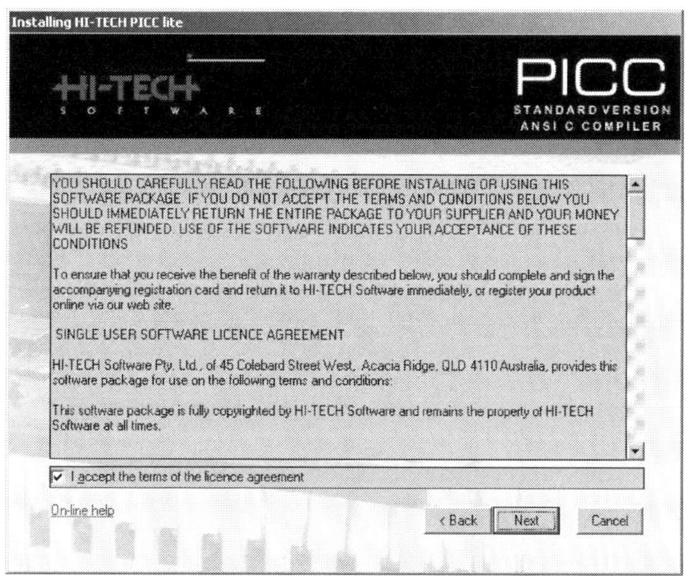

Figure 7-10: HI-TECH PICC-Lite Installation Screen 2

The Figure 7-10 screen asks for you to accept their license agreement. Read it over and click "Next" to accept.

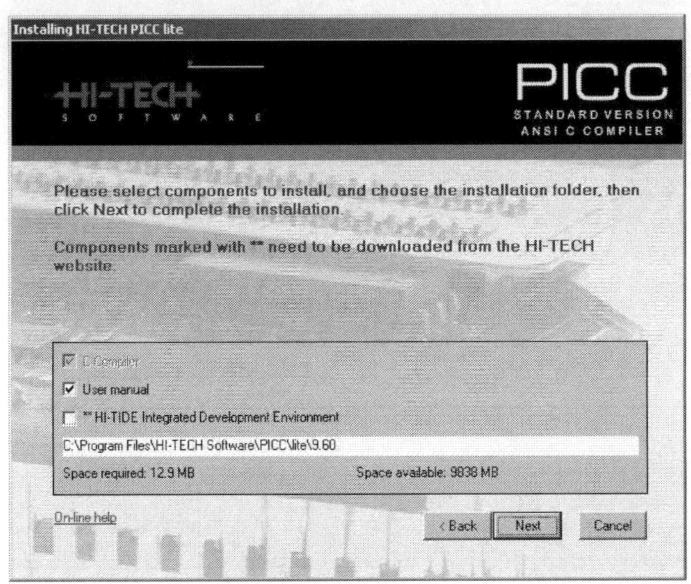

Figure 7-11: HI-TECH PICC-Lite Installation Screen 3

Installation screen three offers a few options. You can download the user manual to your computer which I highly recommend. The other option is HI-TECH's own IDE software for writing the programs. We don't need it since we are using the MPLAB IDE. Notice at the bottom of the choices, it shows the installation path C:\Program Files\HI-TECH Software\PICC\lite|9.60. We will need this later when we setup our first project so I suggest you write this down. Click on "Next" to get the fourth screen shown in Figure 7-12.

Figure 7-12: HI-TECH PICC-Lite Installation Screen 4

The fourth screen offers different languages for the error messages that you choose from the drop down list. There is also a box to check to add the environment path. I highly recommend that you check that box so your computer can find what it needs to compile your programs. Click the "Next" button to move on to screen five.

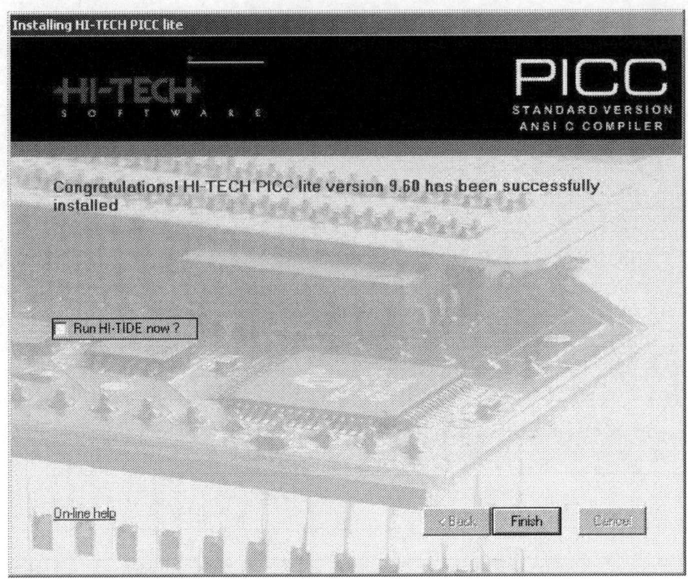

Figure 7-13: HI-TECH PICC-Lite Installation Screen 5

Figure 7-13 shows the final PICC-Lite installation screen. Since we are using MPLAB and didn't install HI-TIDE, I don't suggest you click in the "Run HI-TIDE now ?" box. Just click on the "Finish" button and the installation will jump back to the MPLAB installation. The next screen will be the final MPLAB installation screen shown in Figure 7-14. It will ask if you want to restart your computer and I suggest you click on the "Yes" choice and let your computer reboot with all this new software installed.

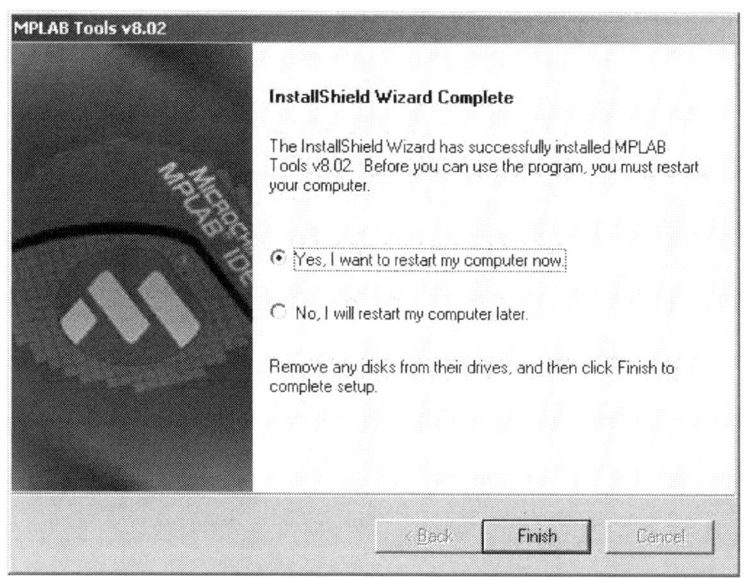

Figure 7-14: MPLAB Final Installation Screen

Project Wizard

After your computer is back up and running, you should have the MPLAB icon on your desktop so double click on it to open MPLAB. You should see the screen in Figure 7-15.

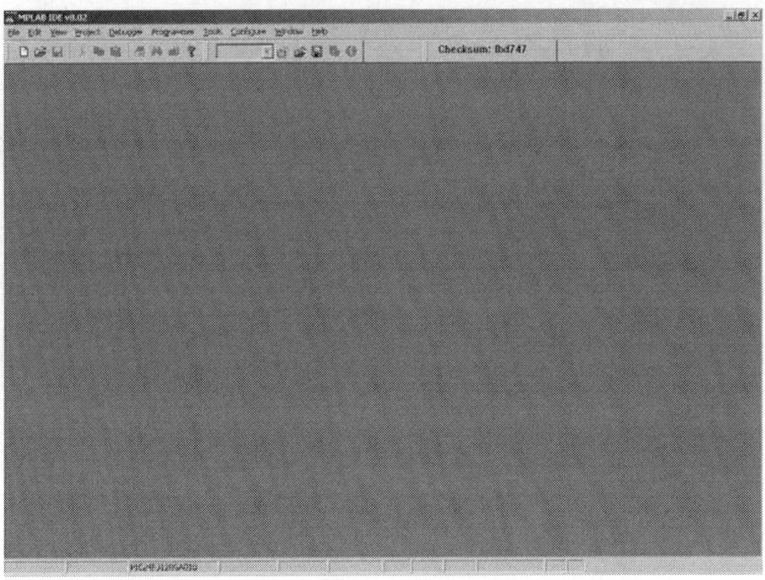

Figure 7-15: MPLAB Initial Screen

As you can see, MPLAB doesn't have much to show but it's where it all starts. I'll now step you through creating a project in MPLAB. Every time you want to start a new software design you will create a project. That project will have all your c files and header files included plus any supporting files you might include. When the project is fully setup, MPLAB will save two files under the suffix ".mcp" and ".mcw" (i.e. project1.mcp).

The ".mcp" file will link all your software files and the ".mcw" will link all the windows you have open in your MPLAB setup. If you open any of the simulator windows or any of the many features built into MPLAB, you won't have to do that again because all of it will be saved in the ".mcw" file so you can easily get back to where you were if you have to shut down your project.

To make creating a project easier, MPLAB has a "wizard" feature that will step you through creating a project. We'll do that here with the first project so you see how to set a project up using the wizard. Figure 7-16 shows the first step.

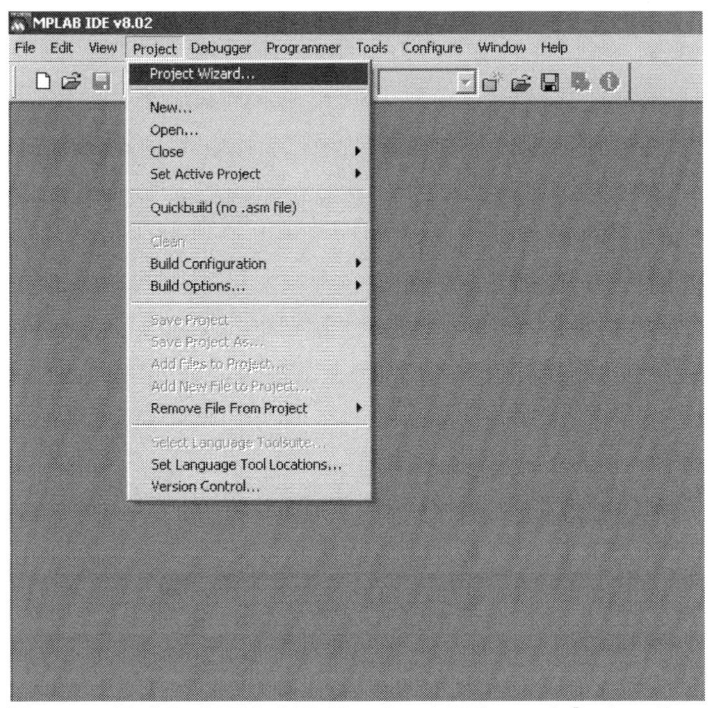

Figure 7-16: MPLAB Project Wizard Selection

Click on the "Project" menu option in the MPLAB screen to get to the Project Wizard option. It will be the first item to select. Click on that Project Wizard sub-menu option to open the window shown in Figure 7-17.

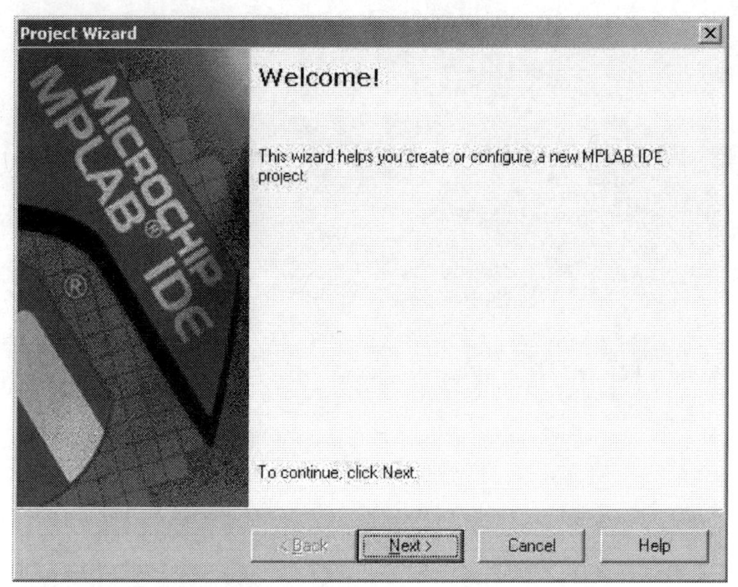

Figure 7-17: MPLAB Project Wizard Screen 1

You should now have a separate window within the MPLAB environment that welcomes you to the project wizard. Click on the "Next" button to move forward.

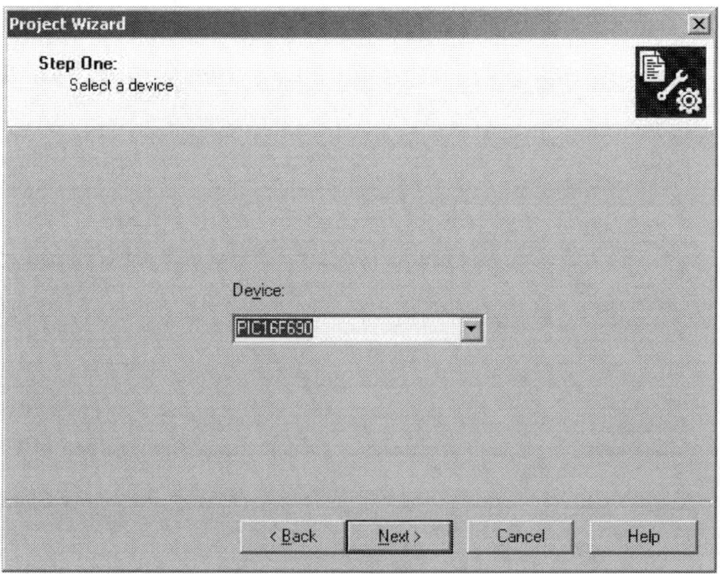

Figure 7-18: MPLAB Project Wizard Screen 2

The step is to pick the microcontroller you plan to design with as shown in Figure 7-18. Throughout this book we will be using the PIC16F690 that comes with the PICkit 2 Starter Kit. Select that from the long list of choices and click the "Next" button to get to the Figure 7-19 screen.

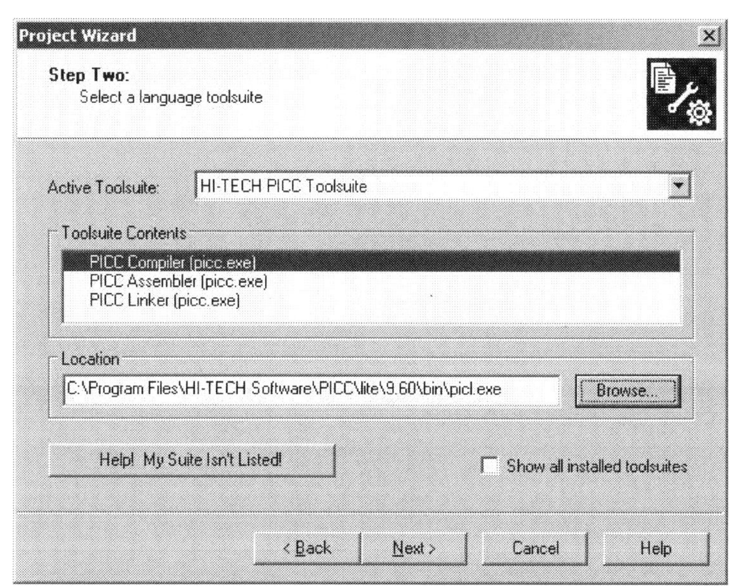

Figure 7-19: MPLAB Project Wizard Screen 3

Now you have to choose the compiler or assembler you want to use and set the location where it is installed. We want to use the PICC-Lite compiler but the list will only show the PICC compiler. The PICC-Lite is a freeware version of the same compiler so go ahead and select HI-TECH PICC Toolsuite from the Active Toolsuite selection window. In the Toolsuite Contents window highlight the PICC Compiler by click on it.

Now the path to the PICC-Lite compiler needs to be set by clicking on the "Browse" button next to the Location window. Refer to the path we wrote down back at Figure 7-11. We are going to point MPLAB to the PICC-Lite C compiler by choosing the "picl.exe" file that is in the "bin" directory of the "\lite\9.60\" subdirectory. This is very important so make sure you have the correct path to the compiler executable file. Click on "Next" to go to the screen in Figure 7-20.

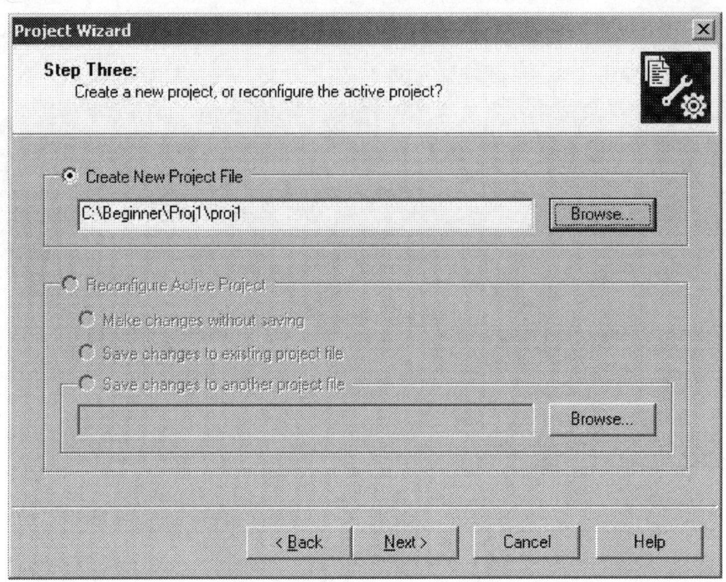

Figure 7-20: MPLAB Project Wizard Screen 4

The next screen is where you create the project name that will also be the prefix of your ".mcp" and ".mcw" files. Click on the "Browse" button to choose where you want the file to be located on your hard drive. Try to choose a path close to the root directory to keep the path name short. MPLAB has a 62 character limit for paths and if you exceed that you won't find out until you compile your first project and get an error. I suggest you just copy the example shown in Figure 7-20. Click "Next" when you are ready to get the screen shown in Figure 7-21.

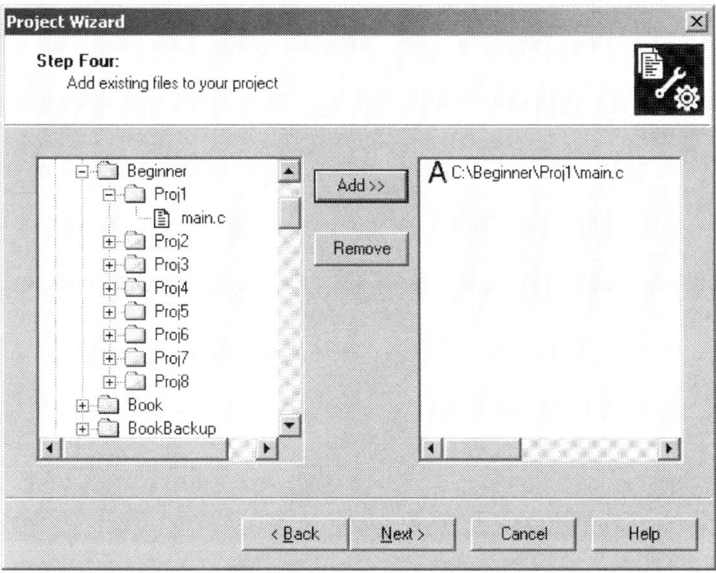

Figure 7-21: MPLAB Project Wizard Screen 5

Step four of the project setup process is where you add any pre-written C files. Typically you will want to modify an existing file to save time rather than start from scratch. In the left window you scroll to find the file you want to modify or load into your project. Highlight that file by clicking on it and then click on the "Add >>" button. It should appear on the right side of the screen.

An optional step is to add the include file for the microcontroller you plan to use. You do that by browsing to the C:\Program Files\Microchip\MPASM Suite directory and load in the P16F690.inc file also. This is the definition file for the PIC16F690 we use throughout this book. The MPLAB compiler will typically find this automatically but I like to keep a copy with my project so I can easily find it to view it. I'm just mentioning this step as an option for reference. I don't show it in the pictures since it's not required.

Next to the selected file you'll see a big "A". If you click on this several times you will see a "C" appear next to it. The "C" will force a copy of the file to be saved in the directory you created back at step three. Some may not recommend this but I like to do this so I keep all my changes in one directory and leave the original files intact. Figure 7-22 shows the final result. Click on the "Next" button to move on from here.

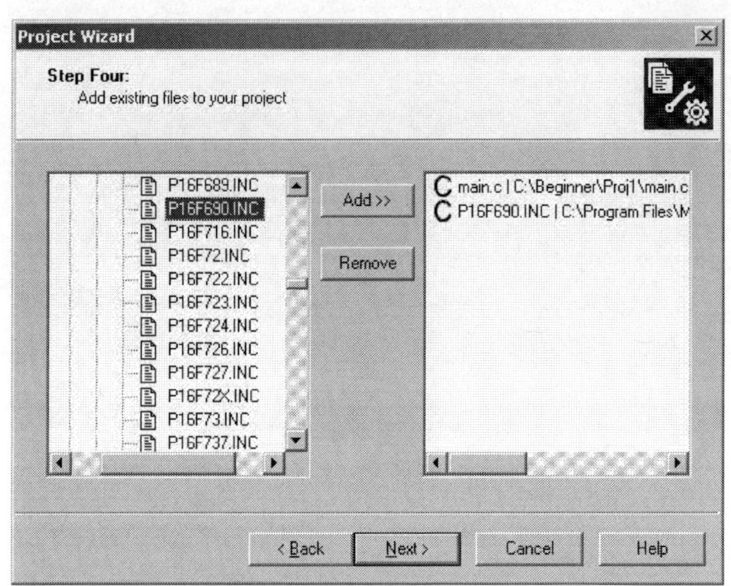

Figure 7-22: MPLAB Project Wizard Screen 6

The screen in Figure 7-23 is the final screen in the project setup. When you get here you have successfully created your first project. Just click finish and the MPLAB window will be the next screen you see as in Figure 7-24.

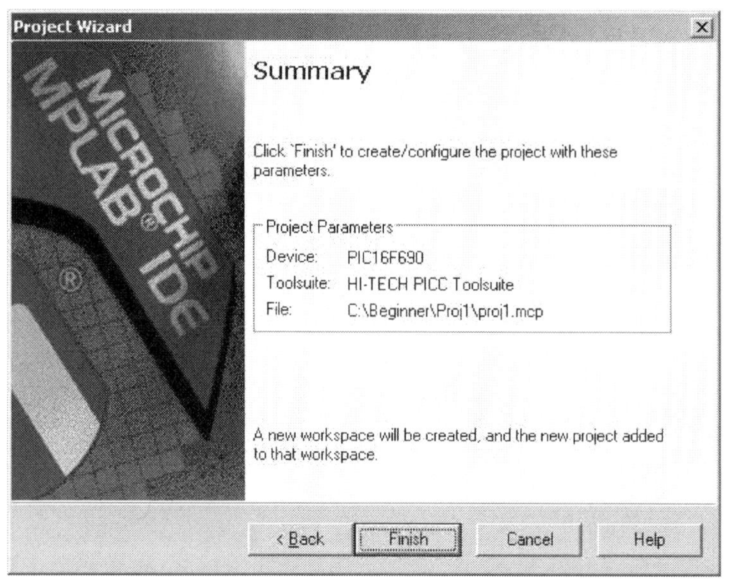

Figure 7-23: Final MPLAB Project Wizard Screen

The project screen shown in Figure 7-24 will initially be blank so you have to open a couple windows. Click on the View>Project option from the top menu and you will see a check mark appear and the project window will also pop up typically in the upper left corner with your project files shown.

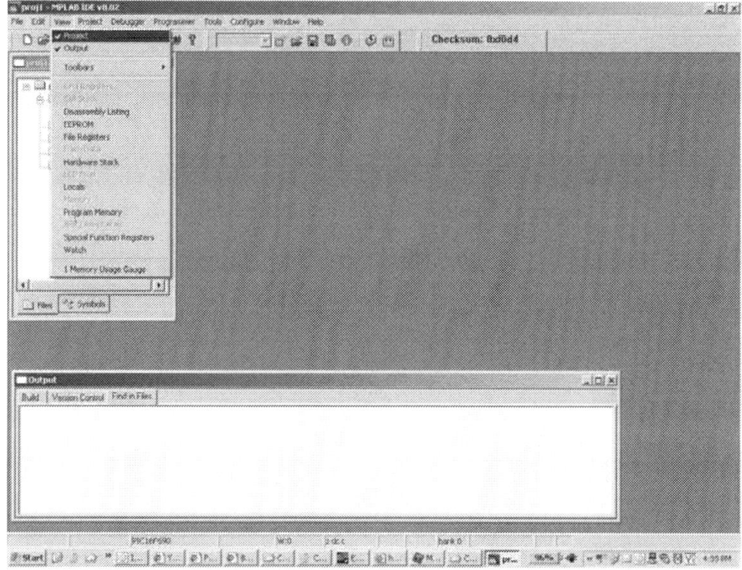

Figure 7-24: MPLAB Project Screens

Click on the View>Output option in the menu and the output window will appear typically at the bottom of the screen. The project window will show the C file you selected earlier. Double click on the C file in your project and the MPLAB Editor window should open up with the code listing displayed. Figure 7-25 shows the editor screen open with the main.c file open.

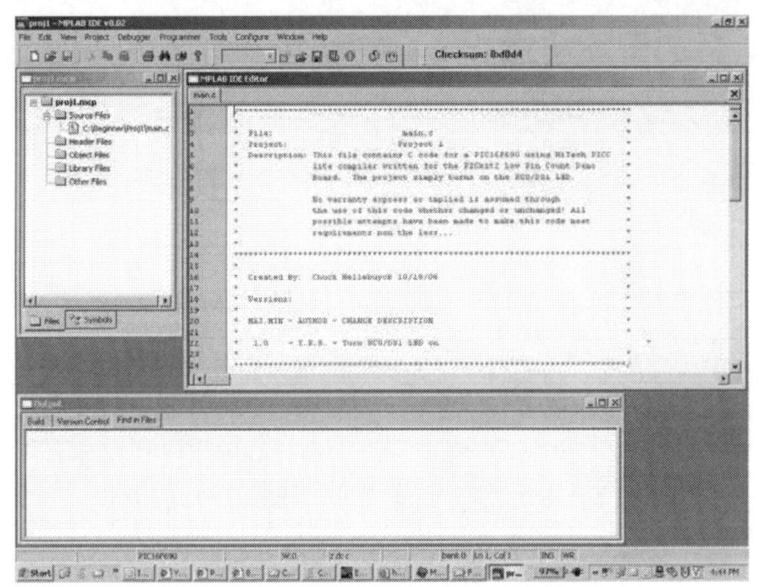

Figure 7-25: MPLAB Editor Window

You would typically modify the file in the editor window to get the program written the way you want. When the file modifications are complete then you are ready to compile it using the PICC-Lite compiler. The MPLAB makes this easy by a simple click of the “Build All” button located at the top of the screen as shown in Figure 7-26.

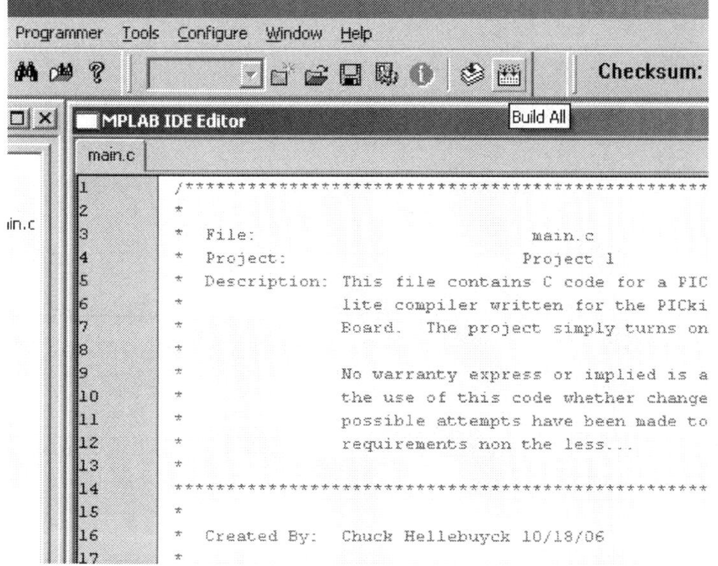

Figure 7-26: MPLAB Project Build All

If your code has any errors this is the step where they show up. The results from the "Build All" will be displayed in the output window. I did it for my sample project and the build was successful as shown in Figure 7-27.

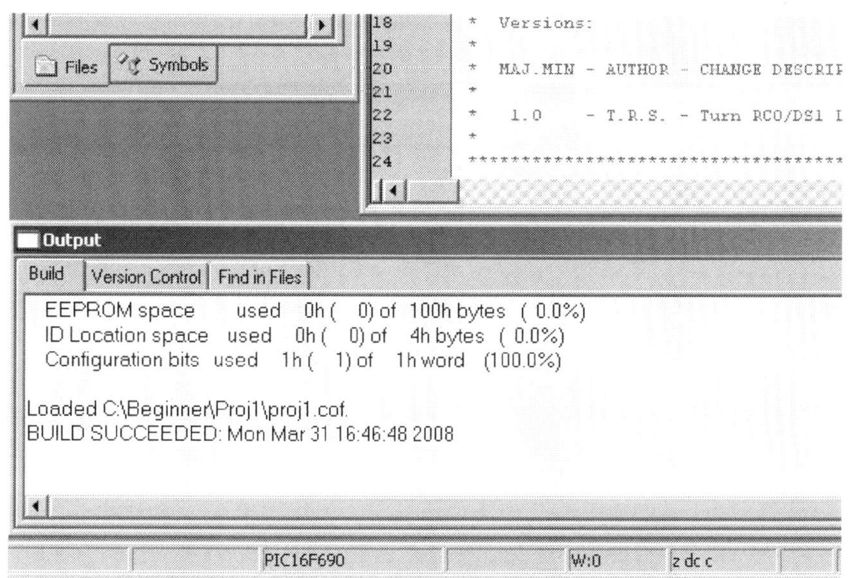

Figure 7-27: MPLAB Build All Success

Once the build is successful we want to load the code into the PIC16F690 and we do that with the PICkit 2 Starter Kit mentioned earlier. Connect the PICkit 2 Starter Kit to your PC USB port and then click on the Programmer>Select Programmer>PICkit 2 menu selection to enable the PICkit 2 as shown in Figure 7-28.

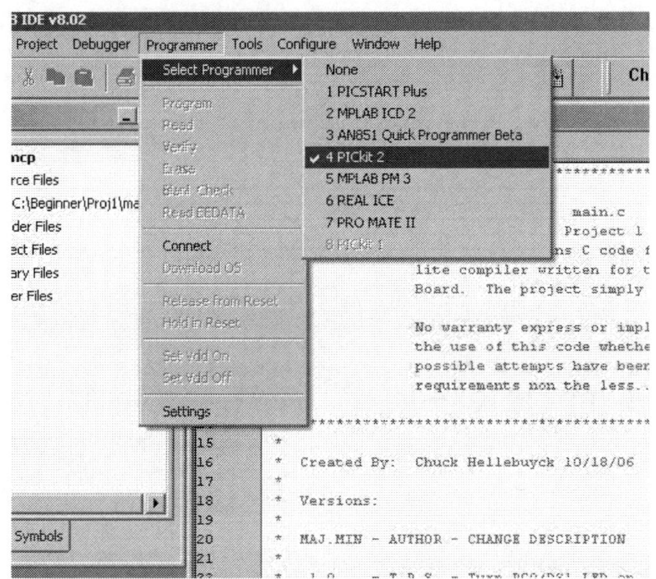

Figure 7-28: Launching PICkit 2 Programmer

The PICkit 2 will respond to signals sent from MPLAB and a status message will be displayed in the PICkit 2 window which is just another tab on the output window as seen in Figure 7-29.

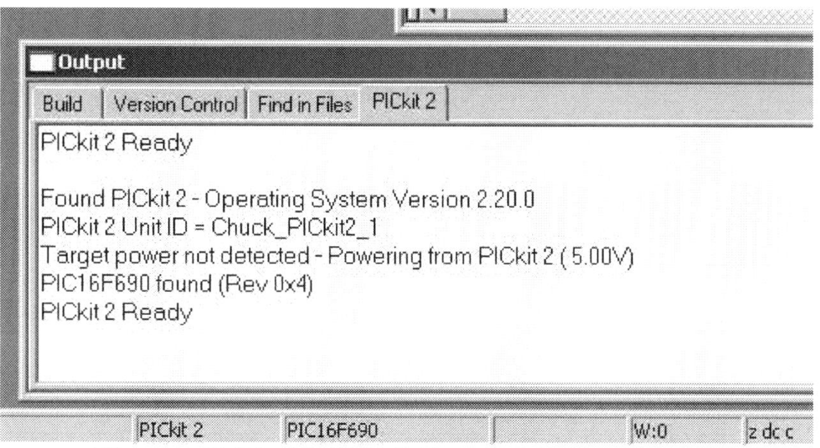

Figure 7-29: PICkit 2 Programmer Ready

The upper corner of the MPLAB window will show the control buttons for the PICkit 2 programmer. The first one with a yellow color and a curved arrow pointing downward is the programming button as seen in Figure 7-30. Make sure the PIC16F690 is in the PICkit 2 Development board the proper way and press the programming button.

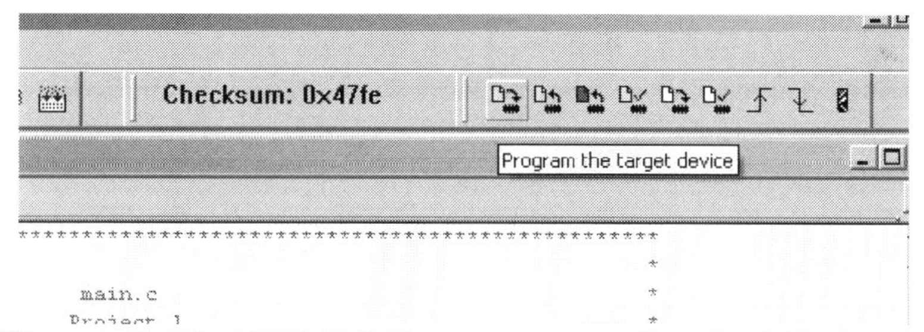

Figure 7-30: PICkit 2 Programmer Control Buttons

The output window will show the results of the programming process which includes erasing the part, then programming it. It's followed by a verification step to compare what is actually in the chip after programming with what MPLAB sent to the programmer. If they match the programming was successful and the output window will show that the PICkit 2 is ready to do it again. Figure 7-29 shows that result.

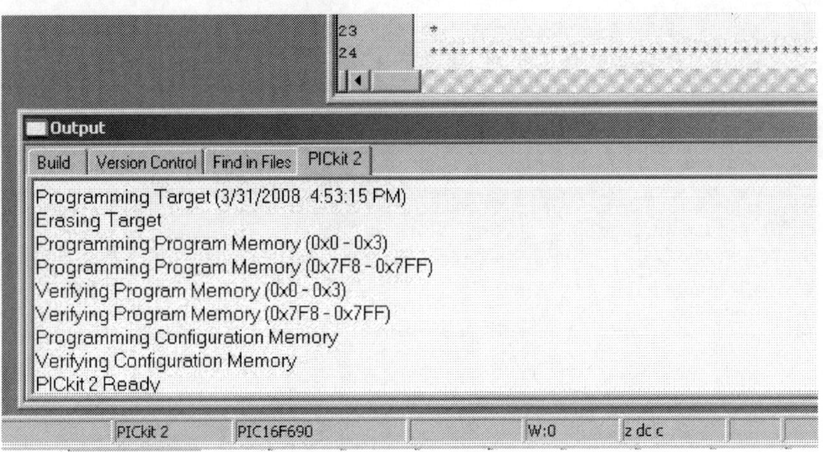

Figure 7-31: PICkit 2 Successfully Programmed the PIC16F690

That is all there is to setting up the development tools to start writing C programs for the PIC16F690. If you used the exact same file I did, then you used the same program that will be covered in the first project chapter. The schematic in Figure 7-32 shows the development board connections. If everything is successful, the DS1 LED will be flashing on the PICkit 2 development board. If it's not flashing the programmer may be holding the chip in reset mode. Click on the release from reset button on the PICkit 2 control buttons which is the rising edge symbol (third symbol from the right). Now let's cover that project in more detail as we enter the project section of this book.

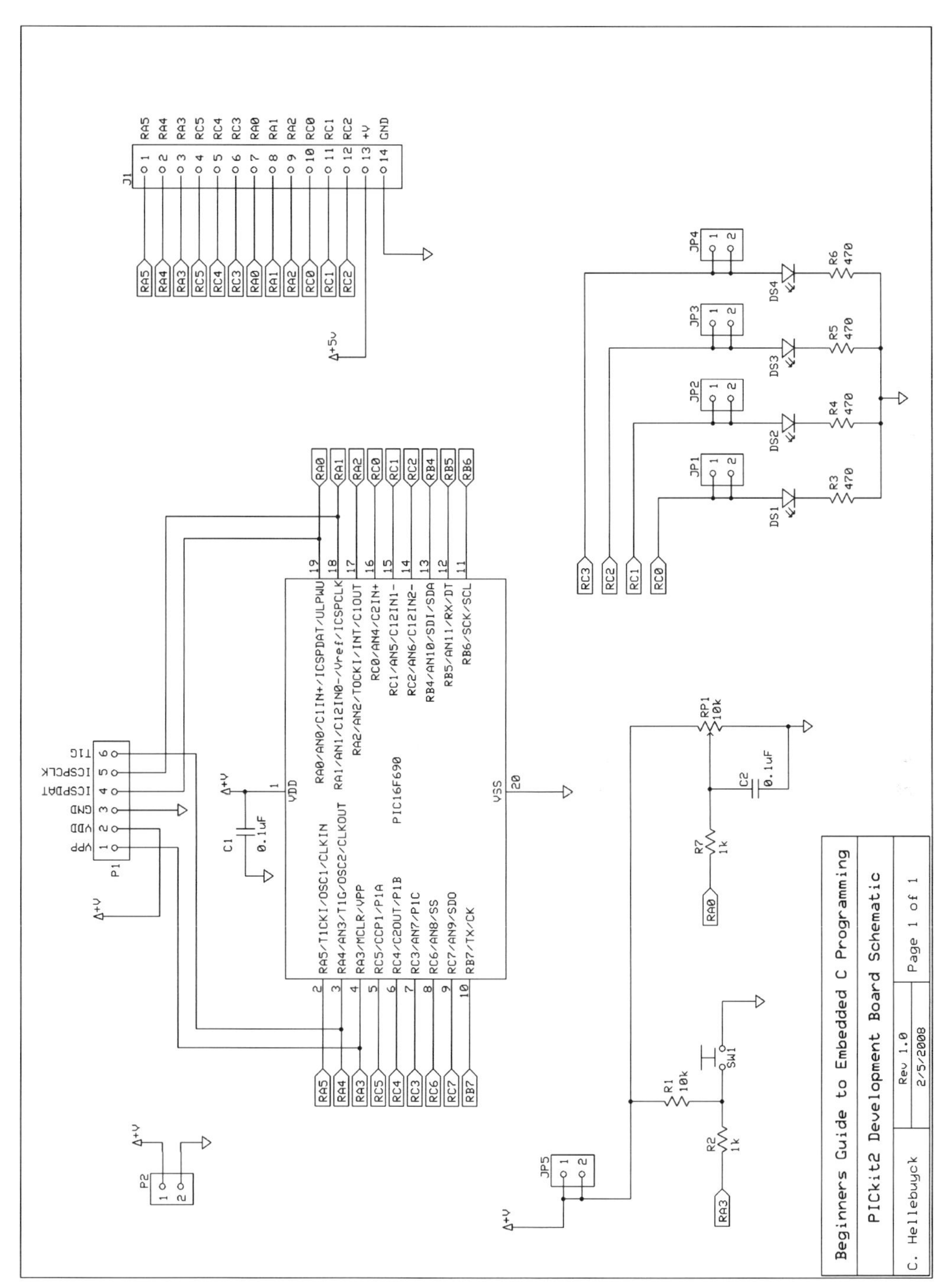

Figure 7-30: PICkit 2 Development Board Schematic

Chapter 8 – Project 1: While Loop

Description

We start off simple and demonstrate how to use the *while* statement to simply light an LED. The LED is wired so a high level on the output port will light the LED and a low will turn it off. This is not a very useful project for performing any function but what this project does is introduce all the pieces that make up a project. If you successfully program the micro and get the LED to light you've accomplished many other things as well; 1) you've setup the MPLAB properly, 2) you've written and compiled the code successfully meaning the PICC-Lite compiler is working and 3) you've verified the PICkit 2 and the PIC16F690 on the development board are communicating properly. Finally, you've gained the confidence to move on to bigger programs knowing all these preliminary steps were successful.

Project Setup

The project setup will not use all the features of the development board. The picture in Figure 8-1 shows the key components used in this project, namely the DS1 LED. The connections between the LED and the micro are already on the board.

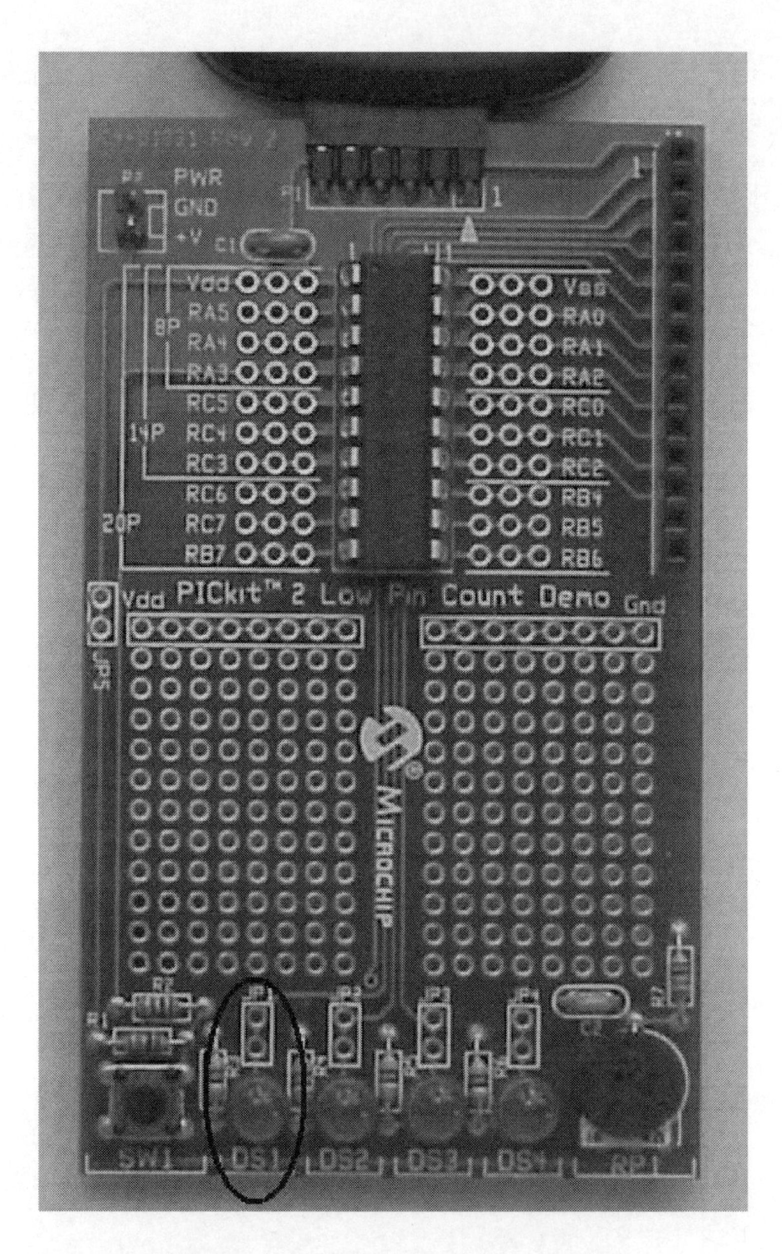

Figure 8-1: Project 1 Hardware

The project 1 schematic (Figure 8-2) shows the connections. The LED is grounded on the board so a high signal from the RC0 pin will light it and a low signal will turn it off. The resistor limits the current. The whole circuit is powered from the PICkit 2 via the USB port of the computer so overall the current limit of the PICkit 2 is about 50 milliamps. The PICkit 2 will power the circuit with 5 volts. Therefore the output of RC0 when set to high will be 5 volts. The red LED will drop about 1.8 volts so the current can be calculated per the simple formula below.

$$(5v - 1.8v) / 470 = 6.8 \text{ milliamps}$$

The I/O pins of the PIC16F690 can drive up to 20 milliamps each so this is well within the limits of the microcontroller. Because all the connections are made we just need to write the software to drive the RC0 pin high to light the LED.

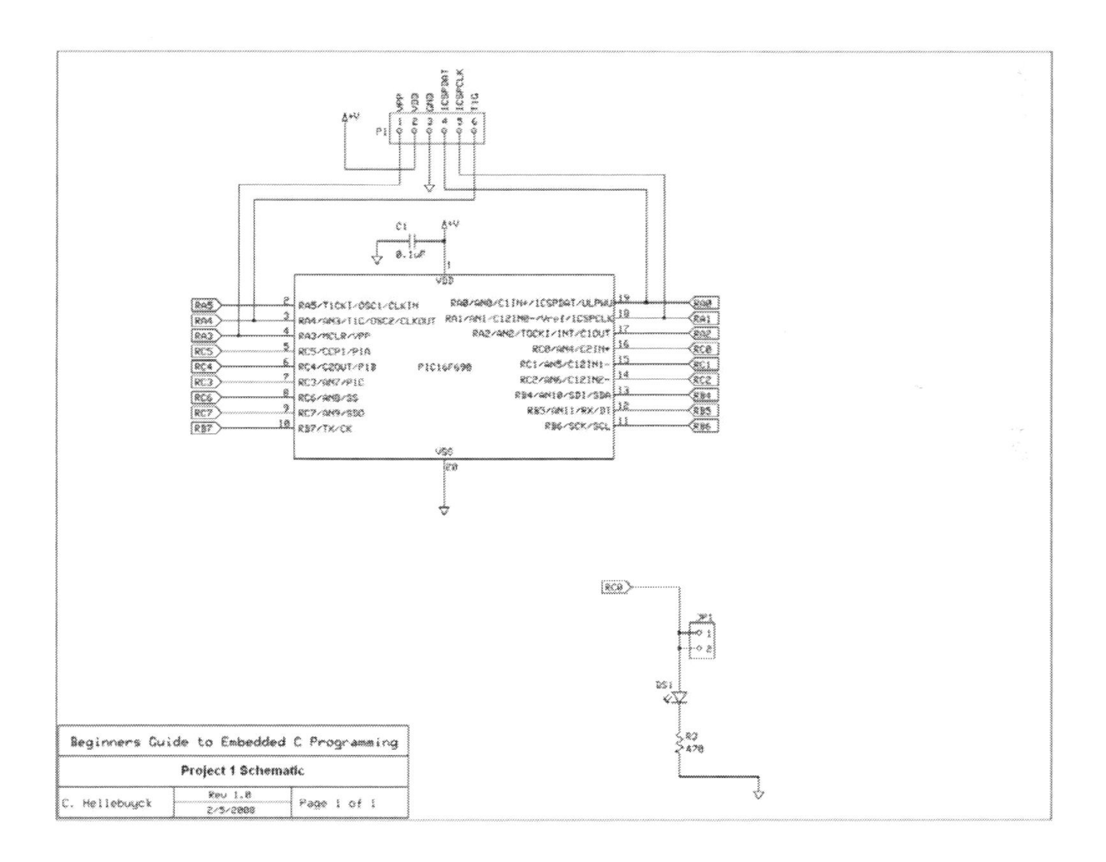

Figure 8-2: Project 1 Schematic

Software

The software is listed below and is quite simple. Most of the space below is taken up by the header. It's a good practice to describe the software with a header so you can easily look back on your description to see what the overall intent of the program is. Sometimes C programs can include several different files all combined at the end. This is known as linking that will be covered in a later project. Having a description of each helps someone else understand what exactly this program is trying to do. In the next section I'll explain how the code works.

```
/*****************************************************************
*                                                                *
* File:            main.c                                        *
* Project:         Project 1                                     *
* Description: This file contains C code for a PIC16F690 using HiTech PICC *
*        lite compiler written for the PICkit 2 Low Pin Count Demo *
*        Board.   The project simply turns on the RC0/DS1 LED.   *
*                                                                *
*                                                                *
*      No warranty express or implied is assumed through         *
*        the use of this code whether changed or unchanged! All  *
*        possible attempts have been made to make this code meet *
*        requirements non the less...                            *
*                                                                *
******************************************************************
*                                                                *
* Created By:  Chuck Hellebuyck 10/18/06                         *
*                                                                *
* Versions:                                                      *
*                                                                *
* MAJ.MIN - AUTHOR - CHANGE DESCRIPTION                          *
*                                                                *
*  1.0   - T.R.S. - Turn RC0/DS1 LED on                          *
*                                                                *
*****************************************************************/

#include <pic.h>      // Include HITECH CC header file


__CONFIG (INTIO & WDTDIS & MCLRDIS & UNPROTECT );
//Internal clock, Watchdog off, MCLR off, Code Unprotected
```

```
main()
{
PORTC = 0x00;       //Clear PortC port
TRISC = 0x00;       //All PortC I/O outputs


        while(1==1)     //loop forever
        {
        RC0 = 1;        // Turn on RC0 LED
        }  //End while

}  //end main
```

How It Works

I'll break the software up into sections so I can explain it easier. The first section is the header. Notice how it starts with a backslash and a star "/*". This signals the C compiler to ignore everything following the slash-star as comments. I then describe everything and even place stars on the edges to make it look like a block. At the very end of the blocks is a star and a backslash "*/". This indicates to the compiler that the comment section is complete.

```
/*****************************************************************
*                                                                 *
* File:           main.c                                          *
* Project:        Project 1                                       *
* Description: This file contains C code for a PIC16F690 using HiTech PICC   *
*             lite compiler written for the PICkit 2 Low Pin Count Demo     *
*             Board.  The project simply turns on the RC0/DS1 LED.          *
*                                                                 *
*                                                                 *
*                                                                 *
*                                                                 *
******************************************************************
*                                                                 *
* Created By:  Chuck Hellebuyck 10/18/06                          *
*                                                                 *
* Versions:                                                       *
```

```
*                                                                   *
*  MAJ.MIN - AUTHOR - CHANGE DESCRIPTION                            *
*                                                                   *
*                                                                   *
*                                                                   *
*********************************************************************/
```

After the header is set the code begins. The first step is to include the header file that the PICC-Lite compiler requires. The #INCLUDE directive handles this. This will be processed before the code is compiled and includes any special setup details the compiler requires.

```
#include <pic.h>      // Include HITECH CC header file
```

The pic.h file is included with the PICC-Lite installation and will be in the "include" directory. This file has lots of specific information about the microcontroller the compiler requires. One of the key lines included in this file is shown below.

```
#if defined(_16F631) || defined(_16F677)    || defined(_16F685)||\
   defined(_16F687)  || defined(_16F689)    || defined(_16F690)
        #include        <pic16f685.h>
#endif
```

This is very important because it looks at what processor you selected in the MPLAB setup and then determines which file to include for all the microcontroller specific details. In our case we are using the PIC16F690 and this section of the "pic.h" file tells the compiler to include the "pic16f685.h" file for all the PIC16F690 specific details. It lists several different parts that can use this file because they all share the same architecture.

You can easily see how one #include command can also #include another file that you don't even see. This can be very confusing to the beginner and it helps to read through the C compiler's manual on these specifics. Each C compiler will handle this a little differently.

The next section of the program is specific to the PIC16F690. There are setup fuses in the microcontroller that configure how you want it to work. For example, if you want to use the built-in internal oscillator as the system clock then you have to set those bits in the configuration register of the PIC16F690. There are several options for the PIC16F690 and those are covered in this books appendix.

The configuration is set with the "_CONFIG" command. All the settings are then placed in parenthesis. In the example below, the internal oscillator is used (INTIO) with the external oscillator pins OSC1 and OSC2 pins used as I/O. The Watchdog Timer is disabled (WDTDIS) and the Master Clear reset pin is also disabled and used as a digital pin (MCLRDIS). Finally the software will be loaded into the PIC16F690 but won't be protected (UNPROTECT). This means that the code can be read from the chip and copied. Since this software is not anything special we don't need to worry about protecting it.

```
__CONFIG (INTIO & WDTDIS & MCLRDIS & UNPROTECT );
//Internal clock, Watchdog off, MCLR off, Code Unprotected
```

The comment line that describes the settings follows a double backslash. The double backslash indicates this line is a comment line. If you are wondering where all these special configuration settings are listed you can read about them in the data sheet for the PIC16F690 that you can download from the Microchip.com website. The actual names used though are defined in that header file "pic16f685.h" that was automatically included with the original #include pic.h command line.

If you open up that pic16f685.h file you will see all the configuration fuse names listed at the bottom of the file. To open that file you will have to find out where the PICC-Lite compiler files are installed. Typically they are under the C:\Program Files\HI-TECH\include directory. I've included that header file in the appendix to make it easier to reference all the nicknames used by the compiler.

Now we enter the section of software where the main loop of code is written. In fact everything prior will be common throughout the projects so if you are feeling a bit confused about everything covered so far, you can feel a bit more relieved knowing that the rest of the program is just the C code.

The main loop is defined by the "main()" statement. The curly brackets contain the specific control statements of the main loop.

```
main()
{
```

The first command statement sets the state of all the PORTC pins. We are only interested in the first PORTC pin or RC0 but to be safe I set all the pins to a zero with this command. This guarantees all the outputs are at a low level which will make all the LEDs off.

```
PORTC = 0x00;        //Clear PortC port
```

The state of PORTC though is not seen at the actual pins yet. You see the PIC16F690 defaults to having all the I/O pins in high impedance input mode. To make the pin an output we have to set the direction register associated with PORTC to an output. The register that controls this is the TRISC register and this name, along with the PORTC name, are both defined in that 16f685.h file.

```
TRISC = 0x00;        //All PortC I/O outputs
```

In reality I am setting all the TRISC bits to zero which will make all of PORTC pins into outputs. I didn't need to change them all just the first pin. Setting TRISC = 0xFE or TRISC = 0b11111110 would set only the RC0 pin to an output. I prefer to know the state of the pins so making them low and outputs allows me to control them. If they are an input, then I really don't know what signal is at the pin without measuring it. On the PIC16F690 a zero in the TRIS register sets it to an output and a one sets it to an input.

Now that we have setup the port, we enter into a *while* loop within the main loop. A simple "while (1=1)" establishes the criteria for this loop. Since one will always equal one, everything between the curly brackets will run continuously.

```
while(1==1)        //loop forever
        {
```

The only action we take in this example is to set the RC0 pin to a high or one. This will light the LED. The label RC0 is also defined in the 16f685.h file automatically included.

```
        RC0 = 1;          // Turn on RC0 LED
```

The program will loop back and do this all again because the *while* command lower curly bracket ends the loop. We are just setting the RC0 pin high multiple times.

```
        }   //End while
```

The main loop curly bracket ends the main loop of code and the program.

```
}   //end main
```

What you cannot see here is the blank line that follows the last curly bracket. The compiler wants to see a blank line or you will get the warning and error shown below.

```
Warning[176] C:\A+Code\C-Book\16F690\Proj1_While_Loop_done\proj1.c 42 : missing newline
Error[139]   : end of file in comment
```

This is definitely an error beginners will get and it can be very confusing. If you look at the two picture below you can see how the first one ends at line 42 which is the same line as the end of the main loop. This will give you this error.

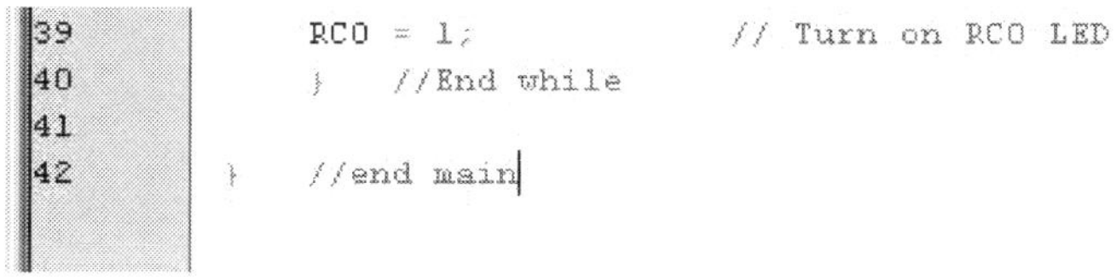

```
39          RC0 = 1;               // Turn on RC0 LED
40          }    //End while
41
42      }   //end main
```

This second picture shows the program ending on a blank line 43. This will build successfully and not give you an error.

```
39          RC0 = 1;             // Turn on RC0 LED
40          }    //End while
41
42    }    //end main
43
```

Next Steps

If you get this far then I assume you have tried out this first project and have successfully built the program and ran it on the PICkit 2 with the development board attached. You can load the file from the project files you downloaded earlier but I recommend you modify it in so you get comfortable writing code. As a next step you can then change the line RC0 = 1 to RC1 = 1 to see if the you can make a different LED light up. Because I wrote the software to make all the PORTC pins outputs, this is an easy one line change.

You could also just add that line so you have both RC0 = 1 and RC1 = 1. This should light both LEDs. That is another option to this simple project.

Chapter 9 – Project 2: Do-While Loop

In this next project I expand on the previous project and make a single LED flash. I maintain the same *while* loop as before and then add a new inner loop using the *do-while* command to demonstrate how to make that statement work for us. The project will use the same hardware setup and control the same LED as Project 1 did but now we'll add some delay and change the state of the LED within the *do-while* loop. The LED is easy to see and understand but it can represent many other things such as a relay that turns an electrical circuit on and off or a buzzer that gives an on/off audible warning. The software for these other types of applications will be the same or at least very similar to flashing an LED so this has more purpose than just a blinking light.

Project Setup

The project setup will use all the same features of the development board used in Project 1. The picture in Figure 9-1 shows the key components used in this project, namely the DS1 LED. The connections between the LED and the micro are already on the board.

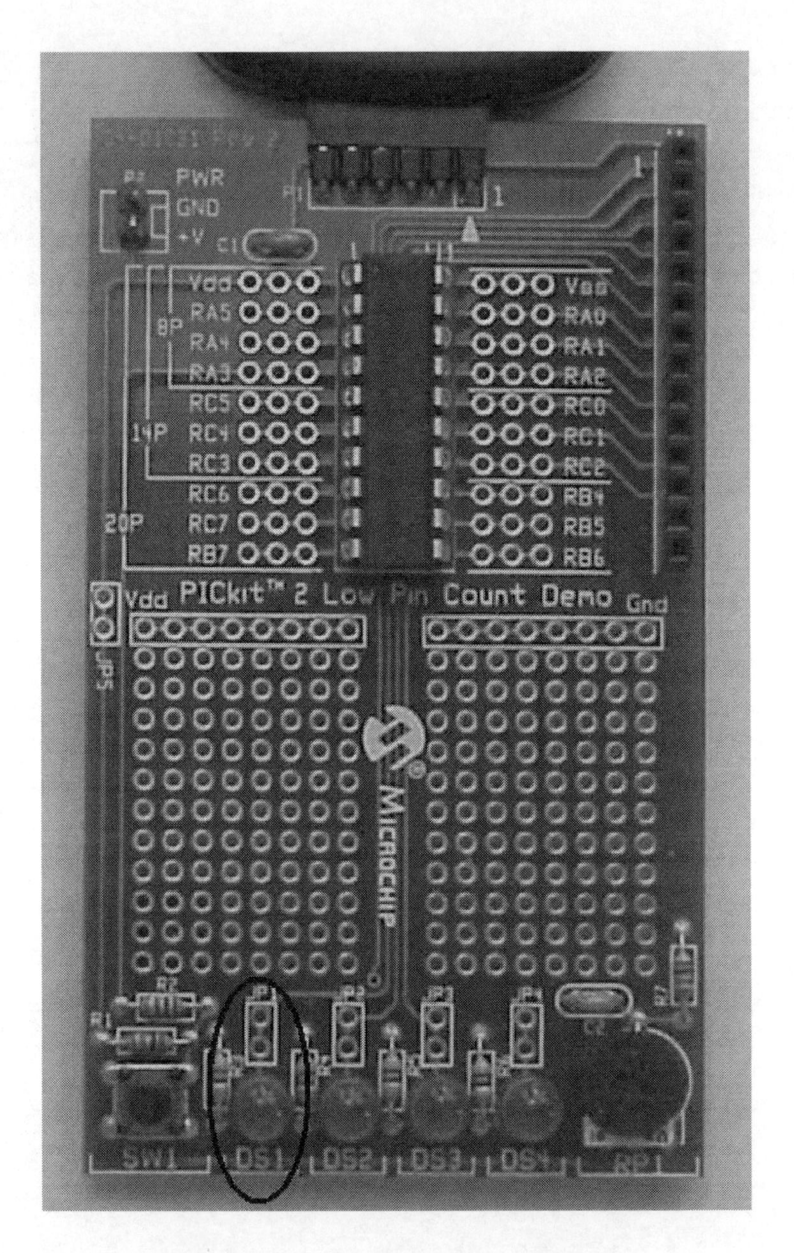

Figure 9-1: Project 2 Hardware

The project 2 schematic (Figure 9-2) shows the exact same connections as project 1. The LED is grounded on the board so a high signal from the RC0 pin will light it and a low signal will turn it off. The resistor limits the current. The whole circuit is powered from the PICkit 2 via the USB port of the computer. Because all the connections are made we just need to write the software to drive the RC0 pin high to light the LED and then low to turn it off to create the blinking action.

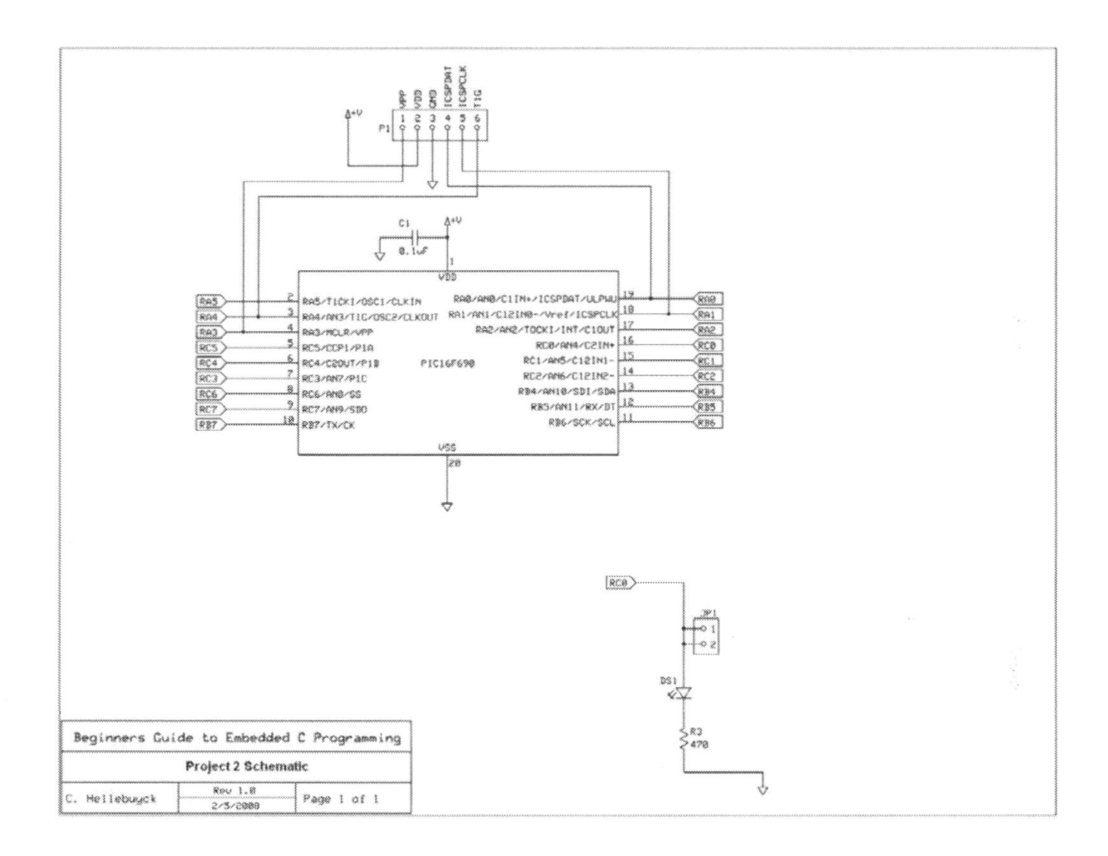

Figure 9-2: Project 2 Schematic

Software

The software is listed below and is a little longer than Project 1. The software has really two sections where one turns the LED on and the other turns it off. A *while* loop is used to control the whole program and then the *do-while* loops are used to change the state of the LED and decrement a variable many times to create a delay. We need the delay to actually see the LED. With the PIC16F690 running on its internal four megahertz oscillator the program would switch the LED on and off so fast the human eye could not detect it. Decrementing a variable many times before changing the state of the LED adds enough delay to allow us to see the LED blink.

```
/*********************************************************************
*                                                                    *
*  File:                          Proj2.c                            *
*  Description: This file contains C code for a PIC16F690 using HiTech*
*               PICC lite compiler written for the PICkit 2 Low Pin  *
*               Count Demo Board. The project demonstrates the       *
*               do-while statement by flashing the RC0/DS1 LED on and *
*               off.                                                 *
**********************************************************************
*
*
*  Created By:  Chuck Hellebuyck 10/18/06
*
*
*
*  Versions:
*
*
*
*  MAJ.MIN - AUTHOR - CHANGE DESCRIPTION
*
*
*********************************************************************/

#include <pic.h>        // Include HITECH CC header file

__CONFIG (INTIO & WDTDIS & MCLRDIS & UNPROTECT );
//Internal clock, Watchdog off, MCLR off, Code Unprotected

unsigned int counter;   // Create delay loop variable with max range of
                        // 0 to 65535

main()
{
PORTC = 0x00;           //Clear PortC port
TRISC = 0x00;           //All PortC I/O outputs


while(1==1)             //loop forever
      {
      RC0 = 1;          // Turn on RC0 LED
      counter = 65535;        // Preset counter to 65535

            do
            {
            counter = counter - 1;  // Decrement count by one
            }
```

```
        while (counter > 0);     // Loop until counter = 0 for delay
                                 // time

    RC0 = 0;                  // Turn off RC0 LED
    counter = 65535;          // Preset counter to 65535

        do
        {
        counter = counter - 1;  // Decrement count by one
        }
        while (counter > 0);     // Loop until counter = 0 for delay
                                 // time

    }     //End while
}     //end main
```

How It Works

I'll jump right to the body of the program since the header is not real interesting and doesn't require anymore explanation then the "/*" "*/" end points.

The same pic.h header file is included which will call in the necessary files to compile the program.

```
#include <pic.h>         // Include HITECH CC header file
```

The configuration is exactly the same for this project and the previous project.

```
__CONFIG (INTIO & WDTDIS & MCLRDIS & UNPROTECT );
//Internal clock, Watchdog off, MCLR off, Code Unprotected
```

Now we enter the unique part of this program. To create the delay required to allow us to see the LED we need to create a variable that we can fill with a value and then decrement it. To get a decent amount of delay I chose a 16 bit variable and declared it unsigned to get the largest possible value. The variable is declared as an unsigned integer with the name "counter". This will allow the variable to contain a value of 0 to 65535.

```
unsigned int counter;    // Create delay loop variable with max range of
                         // 0 to 65535
```

The next part of the program is the main loop that starts with the main label and an open curly bracket. PORTC is also setup the same as Project 1 with all the bits cleared to set all the port pins to a low level and then make them outputs by setting the TRISC register to all zeros.

```
main()
{
PORTC = 0x00;            //Clear PortC port
TRISC = 0x00;            //All PortC I/O outputs
```

The *while* statement is used to create the main functional loop and it starts off with a "1==1" expression to test which will always result in a true state so the loop will run continuously.

```
while(1==1)              //loop forever
      {
```

The first operation on the I/O pins is to set the RC0 pin to a one or high level. This will light the DS1 LED on the development board.

```
      RC0 = 1;                 // Turn on RC0 LED
```

The next line presets the delay variable to its maximum value of 65535. This is done by a simple equation that makes the variable "counter" equal to the literal value 65535 in decimal form.

```
      counter = 65535;         // Preset counter to 65535
```

Now we finally see the *do-while* statement that is used to decrement the "counter" variable in a continuous loop to create the LED delay. Within the *do-while* curly brackets is a simple equation that makes the "counter" variable decrease by one every time the programs loops through.

```
            do
            {
            counter = counter - 1;  // Decrement count by one
            }
```

The program will continue in this loop until the *while* expression is not true. The test is to see if "counter" is greater than zero. When "counter" drops to the value of zero the expression in the *while* statement will no longer be true because zero is not greater than zero. When that occurs, the *do-while* loop is over and the statement that follows the *while* statement will be executed.

```
        while (counter > 0);    // Loop until counter = 0 for delay
                                // time
```

The next statement just sets the RC0 pin to a zero or low level so the LED goes off. Then the "counter" variable is reset to 65535 so another delay can occur.

```
    RC0 = 0;                // Turn off RC0 LED
    counter = 65535;        // Preset counter to 65535
```

The *do-while* loop in this section is identical to the previous *do-while* so the delay is exactly the same.

```
        do
        {
        counter = counter - 1;  // Decrement count by one
        }
        while (counter > 0);    // Loop until counter = 0 for delay
                                // time
```

The closing curly bracket for the original *while* loop is next to indicate the end of that loop.

```
    }     //End while
```

Finally the end of the program is the closing curly bracket for the main loop. Remember to add the blank line after this statement.

```
}     //end main
```

Next Steps

This program can be easily modified to change the amount of the delay to make it flash faster. You can change to a different LED with a couple line changes. You can place a second line next to the statements that light or turn off the LEDs to control a second one. If you set them opposite you can make the LEDs alternate back and forth. For example, in one section make RC0 = 1 and RC1 = 0 and then later make RC0 = 0 and RC1 = 1. This will give you that back and forth LED flash like a train crossing sign. Give that a try.

Chapter 10 – Project 3: Functions

If you tried the "Next Steps" of the last chapter you are already ahead of the game. This project expands on the previous one by flashing two LEDs back and forth like a train crossing signal. The real difference in the project introduced in this chapter is the use of a function. Functions were described earlier in Chapter 6 and this project I use to introduces how to actually use functions.

When driving the LEDs we need to add a delay in the loop to allow the human eye to see them. The microcontroller runs so fast, without a delay the LED will just look like its barely lit. Since we will actually need to call the delay two separate times we save some code space and typing by creating a delay function that we can call from the main loop. If you are familiar with subroutines, this is very similar. The difference here is you can easily pass a parameter to the function. In this project we pass the amount of the delay to the function.

The function created is called "pause" and the value passed to the function represents how many milliseconds to delay. This is a very simple example that shows how easy and useful functions can be. In creating the *pause* function a second function called *msecbase* is also created as the timebase for *pause*.

Project Setup

The project will once again use the LEDs that are pre-wired to the PICkit 2 development board. The DS1 and DS2 LEDs will be flashed back and forth. DS1 LED is connected to the RC0 pin of the PIC16F690 and the DS2 LED is connected to the RC1 pin of the PIC16F690. Figure 10-1 shows the area of the board being used. The schematic is shown in Figure 10-2 and shows the current limiting resistors in series with the LEDs and the connections to the PIC16F690. The same programming connections are used for in-circuit programming and the PICkit 2 will power the development board from the USB port of the PC.

Figure 10-1: Project 3 Hardware

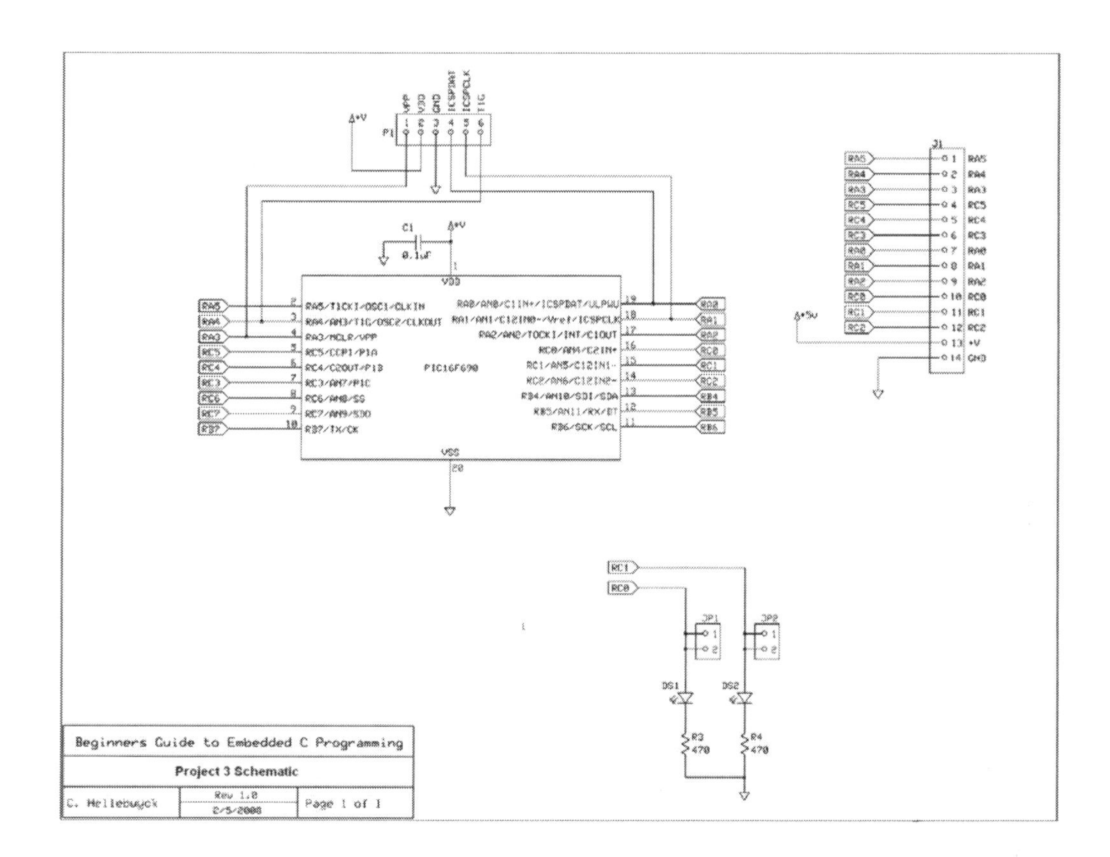

Figure 10-2: Project 3 Schematic

Software

This program actually introduces a couple new items. As mentioned earlier the use of a function to create a delay is used and internal I/O setup is introduced. Many of the I/O pins on the PIC16F690 are multiplexed to other peripherals so you have to turn off some of these peripherals to make the pin just a digital port. It's one of the many small items that can confuse a beginner. The data sheet for the PIC16F690 describes this setup but it's often over-looked by someone just getting started with embedded programming.

I'll explain it more in the "How it Works" section but watch for the following lines as these are controlling some of these multiplexed features.

```
ANSEL = 0;              // Intialize A/D ports off
```

```
CM1CON0 = 0;                    // Initialize Comparator 1 off
CM2CON0 = 0;                    // Initialize Coparator 2 off

/*********************************************************************
*
*
*  File:                          Proj3.C
*
*
*
*  Description: This file contains C code for a PIC16F690 using HiTech
*               PICC lite compiler written for the PICkit 2 Low Pin
*               Count Demo board.
*               The project simply alternates flashing the RC0/DS1 and
*               RC1/DS2 LEDs at a 500 msec rate. This project also
*               demonstrates how to use a Function.
*
*
*
*********************************************************************
*
*
*  Created By:  Chuck Hellebuyck 10/18/06
*
*
*
*  Versions:
*
*
*
*  MAJ.MIN - AUTHOR - CHANGE DESCRIPTION
*
*
*
*
********************************************************************/

#include <pic.h>        // Include HITECH CC header file

/*
PIC16F690 Configuration
*/
__CONFIG (INTIO & WDTDIS & MCLRDIS & UNPROTECT );
//Internal clock, Watchdog off, MCLR off, Code Unprotected

void pause( unsigned short usvalue );//Establish pause routine function
void msecbase( void );          //Establish millisecond base function

main()
{
ANSEL = 0;                      // Intialize A/D ports off
CM1CON0 = 0;                    // Initialize Comparator 1 off
CM2CON0 = 0;                    // Initialize Coparator 2 off
```

```
PORTC = 0x00;                   //Clear PortB port
TRISC = 0x00;                   //All PortB I/O outputs


while(1==1)                     //loop forever
      {
            RC0 = 1;            // Turn on RB0 LED
            RC1 = 0;            // Turn off RB1 LED
            pause(500);         // Delay for .5 seconds

            RC0 = 0;            // Turn off RB0 LED
            RC1 = 1;            // Turn on RB1 LED
            pause(500);         // Delay for .5 seconds

      }     //End while

}     //end main


//*******************************************************
//pause - multiple millisecond delay routine
//*******************************************************

void pause( unsigned short usvalue )
{
      unsigned short x;

      for (x=0; x<=usvalue; x++)//Loop through a delay equal to usvalue
            {                   // in milliseconds.
            msecbase();         //Jump to millisec delay routine
            }
}


/*************************************************************
* msecbase - 1 msec pause routine                            *
* The Internal oscillator is set to 4 Mhz and the            *
* internal instruction clock is 1/4 of the oscillator.       *
* This makes the internal instruction clock 1 Mhz or         *
* 1 usec per clock pulse.                                    *
* Using the 1:4 prescaler on the clock input to Timer0       *
* slows the Timer0 count increment to 1 count/4 usec.        *
* Therefore 250 counts of the Timer0 would make a one        *
* millisecond delay (250 * 4 usec). But there are other      *
* instructions in the delay loop so using the MPLAB          *
* stopwatch we find that we need Timer0 to overflow at       *
* 243 clock ticks. Preset Timer0 to 13 (0D hex) to make      *
* Timer0 overflow at 243 clock ticks (256-13 = 243).         *
* This results in a 1.001 millisecond delay which is         *
* close enough.                                              *
*************************************************************/

void msecbase(void)
{
```

```
        OPTION = 0b00000001;        //Set prescaler to TMRO 1:4
        TMRO = 0xD;                 //Preset TMRO to overflow on 243 counts
        while(!T0IF);               //Stay until TMRO overflow flag equals 1
        T0IF = 0;                   //Clear the TMR0 overflow flag
}
```

How It Works

The program starts off as the others by including the necessary pic.h file. Then the configuration is added. I used the "/*" and "*/" to create a new header for the configuration settings.

```
#include <pic.h>        // Include HITECH CC header file

/*
PIC16F690 Configuration
*/
__CONFIG (INTIO & WDTDIS & MCLRDIS & UNPROTECT );
//Internal clock, Watchdog off, MCLR off, Code Unprotected
```

This next section is added strictly for the function. Actually the function "pause" includes a second function called "msecbase". What I did is create a loop that is very close to a one millisecond delay and then I use that as the time base for the "pause" function. I'll show more later but to use these functions I could put them at the top of the program or place them after the main loop of code. I chose to put them after the main loop so the compiler needs to know they exist prior to compiling the main loop. This is done with a function prototype placed before the main loop. It's just the first line of the function followed by a semi-colon.

The pause function and the msecbase function both operate without returning a value so they start with a void. The pause function will accept a delay value from the main loop so a variable to hold that value in the pause function needs to be created. This is shown in the parenthesis as an unsigned short variable named "usvalue". This will accept values from 0 to 65535. These are the function prototypes.

```
void pause( unsigned short usvalue );//Establish pause routine function
void msecbase( void );         //Establish millisecond base function
```

Now that the functions are defined the main loop of code is entered. At the top of the main loop are three statements that were not in any of the previous projects but probably should have been. If you look at the schematic for the PICkit 2 development board you will see that the PIC16F690 has multiple names for the various I/O pins. This is because those pins are connected to multiple circuits inside. Those connections are controlled by internal register bits that you can set or clear in software.

The Analog to Digital (A/D) converter is often shared with many of the digital I/O pins and they take on the name ANx where x is the A/D port number. There are some pins that also share the connection with the internal comparators shown as a C1 or C2 variant. We want to use the RC0 and RC1 pins as digital pins so to make sure the comparators and A/D converter is turned off we have to set the corresponding bits in the control registers to zero.

The ANSEL register controls the A/D ports and that register layout is shown next.

REGISTER 4-3: ANSEL: ANALOG SELECT REGISTER

R/W-1	R/W-1	R/W-1	R/W-1	R/W-1	R/W-1	R/W-1	R/W-1
ANS7	ANS6	ANS5	ANS4	ANS3	ANS2	ANS1	ANS0
bit 7							bit 0

Legend:			
R = Readable bit	W = Writable bit	U = Unimplemented bit, read as '0'	
-n = Value at POR	'1' = Bit is set	'0' = Bit is cleared	x = Bit is unknown

bit 7-0 **ANS<7:0>**: Analog Select bits
Analog select between analog or digital function on pins AN<7:0>, respectively.
1 = Analog input. Pin is assigned as analog input[1].
0 = Digital I/O. Pin is assigned to port or special function.

Note 1: Setting a pin to an analog input automatically disables the digital input circuitry, weak pull-ups and interrupt-on-change if available. The corresponding TRIS bit must be set to Input mode in order to allow external control of the voltage on the pin.

REGISTER 4-4: ANSELH: ANALOG SELECT HIGH REGISTER[2]

U-0	U-0	U-0	U-0	R/W-1	R/W-1	R/W-1	R/W-1
—	—	—	—	ANS11	ANS10	ANS9	ANS8
bit 7							bit 0

Legend:			
R = Readable bit	W = Writable bit	U = Unimplemented bit, read as '0'	
-n = Value at POR	'1' = Bit is set	'0' = Bit is cleared	x = Bit is unknown

bit 7-4 **Unimplemented:** Read as '0'
bit 3-0 **ANS<11:8>**: Analog Select bits
Analog select between analog or digital function on pins AN<7:0>, respectively.
1 = Analog input. Pin is assigned as analog input[1].
0 = Digital I/O. Pin is assigned to port or special function.

Note 1: Setting a pin to an analog input automatically disables the digital input circuitry, weak pull-ups and interrupt-on-change if available. The corresponding TRIS bit must be set to Input mode in order to allow external control of the voltage on the pin.
2: PIC16F677/PIC16F685/PIC16F687/PIC16F689/PIC16F690 only.

There are actually two registers for all the A/D pins so there is an ANSEL and an ANSELH. The lower 8 pins which are AN0 thru AN7 are controlled by the ANSEL register and that is all this project really cares about so I only modify this register. You can see by the description of the bits in the register that a "1" makes the pin an analog or A/D pin while a "0" makes it a digital pin. I only needed to set the AN4 and AN5 pins to digital since they share the RC0 and RC1 ports. To make things easier though I just set all the A/D pins controlled by ANSEL to digital by making the register equal to zero.

```
main()
{
ANSEL = 0;                          // Intialize A/D ports off
```

The comparators can be enabled or disabled by setting or clearing bit 7 of the CM1CON0 or CM1CON1 registers. If you look at all the options in the comparator registers you have a lot of options to control when using the comparators. To make it simple I just shut them off and set all the bits to zero. This disables both comparators.

```
CM1CON0 = 0;                        // Initialize Comparator 1 off
CM2CON0 = 0;                        // Initialize Comparator 2 off
```

You can see that understanding all the features of a particular microcontroller can be very helpful to writing embedded C code. Some parts default to the A/D pins on and some have the comparator turned on. The PIC16F690 defaults to having the A/D turned on when the pin is set to an input. We used them as an output in the previous projects so we got away with not pre-setting these registers. I purposely did that to eliminate confusion but in future projects you should always preset these registers.

Note:
These register names along with the port register names are defined in pic.h as all capitals. You have to keep them all capitals to prevent errors.

The next set of registers we set is the same as previous projects. We set the digital ports pins of PORTC to zero and then set the state to all outputs by setting the TRISC register to all zeros.

```
PORTC = 0x00;                       //Clear PortB port
TRISC = 0x00;                       //All PortB I/O outputs
```

A *while* statement is used to form the main loop of code that will repeat.

```
while(1==1)                         //loop forever
      {
```

The first block of code sets the RC0 pin high and the RC1 pin low. This turns the DS1 LED on and the DS2 LED off.

```
        RC0 = 1;              // Turn on RB0 LED
        RC1 = 0;              // Turn off RB1 LED
```

Then the function *pause* is called and the value of 500 is passed to the function. This will create a 500 millisecond delay or 0.5 seconds. This will allow us to see the state of the LEDs.

```
        pause(500);           // Delay for .5 seconds
```

The LEDs are then reversed making DS1 off and DS2 on. Another 500 millisecond delay is called.

```
        RC0 = 0;              // Turn off RB0 LED
        RC1 = 1;              // Turn on RB1 LED
        pause(500);           // Delay for .5 seconds
```

This is the end of the loop and this will continue over and over until one does not equal one as stated in the *while* statement (one will always equal one so this lasts forever)

```
    }       //End while

}       //end main
```

After the closing curly bracket of the main loop is where the functions are placed. If these were at the top then the prototypes we declared earlier would not be necessary.

```
//******************************************************
//pause - multiple millisecond delay routine
//******************************************************
```

After a short title block the prototype statement appears again but without the semi-colon at the end. The *pause* function first creates a counting variable called “x”. This is only visible to the function so this is a local variable.

```
void pause( unsigned short usvalue )
{
      unsigned short x;
```

Here I jump ahead a bit and use a *for* statement to create the delay based on the value passed to the function. I'll cover more about the *for* statement in the next chapter project but its easy to understand. The value of *x* is preset to zero and then it is tested to see if it is greater than the variable *usvalue* which was loaded with 500 when the function was called by the main loop.
I violate my own rule about not using shortcuts by using an x++ statement which is the same as x = x + 1. This increments the value of x and then it is tested against the *usvalue* variable. Before doing that though, the function *msecbase* is called. I'll cover that next but this is just a fixed time delay of 1 millisecond. If we jump to *msecbase* 500 times before returning, then we will have created a 500 millisecond delay.

```
        for (x=0; x<=usvalue; x++)//Loop through a delay equal to usvalue
              {                    // in milliseconds.
              msecbase();          //Jump to millisec delay routine
              }
}
```

The *pause* function ends with a curly bracket.

In all the calculations of time you might notice that the extra instruction clock pulses needed to drive the *for* statement are not included in this delay routine. You are correct to catch that because I did not take that into account. This is not the perfect calculation of an accurate time base but it's very close so I went with it.

The next function is the *msecbase* routine that I did put some effort into making it accurate. In this case I actually use one of the internal timers of the PIC16F690 to create the time base. In the title block I try to describe how I came up with the timer preload value to get close to an accurate 1 millisecond time base. You can read through that. The stopwatch I refer to is a software tool option within the MPLAB IDE. It's a handy way to measure your code execution. You run it from within the MPLAB simulator.

```
/*************************************************************
* msecbase - 1 msec pause routine                            *
* The Internal oscillator is set to 4 Mhz and the            *
* internal instruction clock is 1/4 of the oscillator.       *
* This makes the internal instruction clock 1 Mhz or         *
* 1 usec per clock pulse.                                    *
* Using the 1:4 prescaler on the clock input to Timer0       *
* slows the Timer0 count increment to 1 count/4 usec.        *
```

```
* Therefore 250 counts of the Timer0 would make a one          *
* millisecond delay (250 * 4 usec). But there are other         *
* instructions in the delay loop so using the MPLAB             *
* stopwatch we find that we need Timer0 to overflow at          *
* 243 clock ticks. Preset Timer0 to 13 (0D hex) to make         *
* Timer0 overflow at 243 clock ticks (256-13 = 243).            *
* This results in a 1.001 millisecond delay which is            *
* close enough.                                                 *
*************************************************************/
```

The prototype for the *msecbase* function shows up again without the semi-colon.

```
void msecbase(void)
{
```

Inside the PIC16F690 are several shift register counters that are driven from the instruction clock. This makes them timers instead of counters. You can control how you want them to increment by controlling a pre-scaler that intercepts the clock signal prior to entering the counter. The option register of the PIC16F690 is where this is set. By setting just the first bit of the OPTION register, the internal clock will tick four times before incrementing TIMER0 one count.

```
      OPTION = 0b00000001;     //Set prescaler to TMRO 1:4
```

The header block described how I calculated the preset for the TIMER0 and this next line shows how that value is loaded into the TMR0 register. You do it by just making the register equal to the value you want to load it with. We want the timer to overflow at 243 counts so we preload the TIMER0 with decimal 13 or hex value 0D.

```
      TMR0 = 0x0D;                     //Preset TMRO to overflow on 243
counts
```

The next line is another use of the *while* statement. The TIMER0 is an 8 bit counter/timer so after it is incremented 256 times it rolls over and resets back to 0. When this occurs the internal circuitry sets a bit in the INTCON register called the T0IF flag. Because I calculated the preset value to make TIMER0 overflow after 1 milliseconds of clock ticks, this loop only has to test that T0IF bit to determine if 1 millisecond has passed. This *while* statement does that by staying at this line until the T0IF bit is not set. This is a one line loop.

```
    while(!T0IF);           //Stay until TMRO overflow flag equals 1
```

After the T0IF bit gets set, we have to clear it for the next time we run the one millisecond delay. That is an easy equal statement.

```
    T0IF = 0;               //Clear the TMR0 overflow flag
}
```

This is all there is to the millisecond delay time base function. It's a little more accurate because it uses the internal timer of PIC16F690. You might have to read the PIC16F690 data sheet to fully understand this but all the timers work the same way so this is a good thing to learn.

Next Steps

There was a lot covered here so my guess is you may be a bit confused about all the concepts introduced. There is no substitute for reading the data sheet when programming embedded applications. One easy modification though is to play with the delay time value and change it from 500. You could put an oscilloscope on the LED pins to actually measure how accurate the delay is.

You could also try and calculate a new time base value for the millisecond time base to make it a half millisecond time base or some other value. If you do this you will better understand how to work with the timers which is something I highly recommend you learn.

Chapter 11 – Project 4: For-Loop

Much of the last project is reused here and expanded on. There are numerous times when I've seen a movie special effect use a series of LEDs to create light scrolling back and forth to represent a computer calculating something. I've also seen science fiction movies have an alien mask with LEDs for eyes that scan an object with that back and forth light. It turns out to be a very easy project to recreate using a *for* loop. I add a *for loop* to the main project code in the previous chapter to light the LED's on the PICkit 2 development board sequentially, one at a time and then reverse it. This creates that back and forth movement of light.

I'll use functions for the pause and time base requirements. This is also not limited to four LEDs as this project shows but you can expand it beyond where this project goes.

Project Setup

The setup uses all four LEDs on the PICkit 2 development board shown in Figure 11-1. The schematic is shown in Figure 11-2. The LEDs are all wired to the same port so scrolling through them is easier because lighting them just requires moving a bit in the PORTC register. If the PICkit 2 board had the LEDs wired to different ports, this project could still be done but it would be a little more complicated. The PICkit 2 powers the development board.

Figure 11-1: Project 4 Hardware

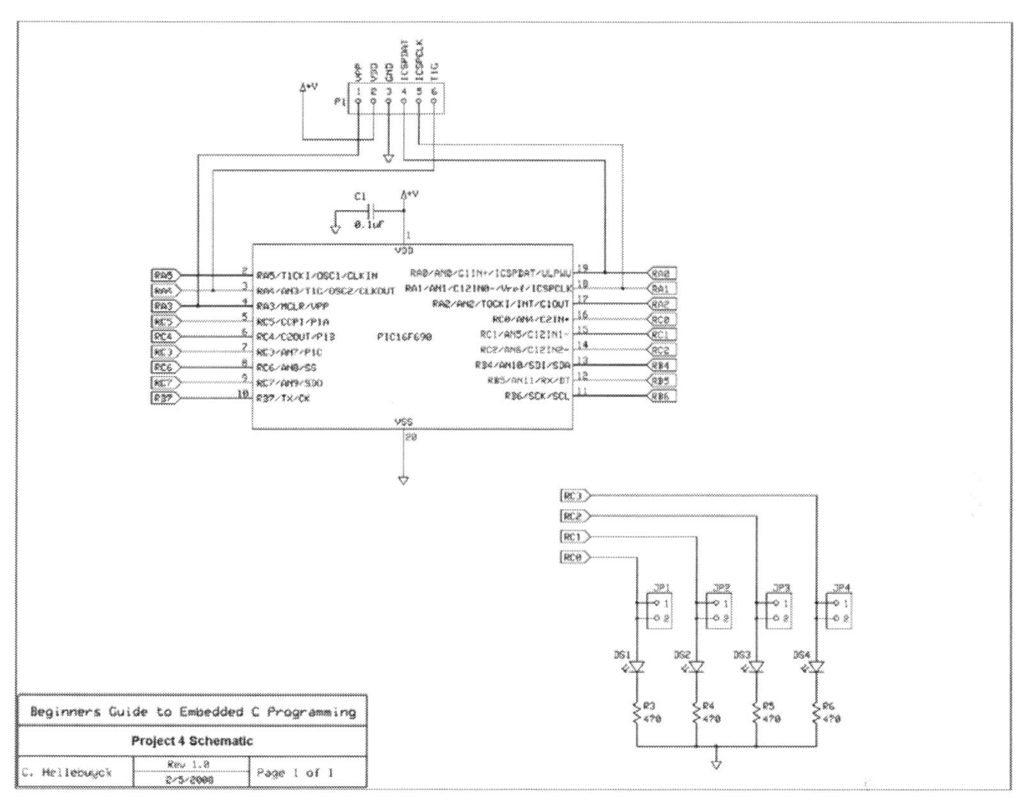

Figure 11-2: Project 4 Schematic

Software

```
/**********************************************************************
*
*
*  File:                          Proj4.C
*
*
*
*  Description: This file contains C code for a PIC16F690 using HiTech
*               PICC lite compiler version 9.60. The code is written
*               for the PICkit 2 low pin count demo board.
*
*               The project scrolls thru the RC0/DS1 to RC3/DS4 LEDs
*               using the For statement and a 500 msec delay Function.
*
*
*  Created By:  Chuck Hellebuyck 10/18/06
*
*
*
```

```
*  Versions:
*
*
*
*  MAJ.MIN - AUTHOR - CHANGE DESCRIPTION
*
*
*
*
*
*********************************************************************/

#include <pic.h>        // Include HITECH CC header file

/*
PIC16F690 Configuration
*/
__CONFIG (INTIO & WDTDIS & MCLRDIS & UNPROTECT );
//Internal clock, Watchdog off, MCLR off, Code Unprotected

unsigned short delay=500; //Initialize on/off delay value to 500 msec

void Pause( unsigned short msvalue ); //Establish pause routine
                                      //function
void msecbase( void );  //Establish millisecond base function

main()
{
ANSEL = 0;                      // Intialize A/D ports off
CM1CON0 = 0;                    // Initialize Comparator 1 off
CM2CON0 = 0;                    // Initialize Coparator 2 off

PORTC = 0x00;                   //Clear PortB port
TRISC = 0x00;                   //All PortB I/O outputs

while(1==1)                     //loop forever
      {
      int x;
      for(x=1; x<9; x=x*2)
            {
            PORTC = x;                // Turn on next LED
            Pause(delay);             // Delay for .5 seconds
            }      // End For
      }      //End while
}      //end main


//*******************************************************
//pause - multiple millisecond delay routine
//*******************************************************

void Pause( unsigned short msvalue )
{
      unsigned short x;
```

```
        for (x=0; x<=msvalue; x++)      //Loop through a delay equal to
              {                         // msvalue in milliseconds.
              msecbase();         //Jump to millisec delay routine
              }
}


//***************************************************
//msecbase - 1 msec pause routine
//***************************************************

void msecbase(void)
{
      OPTION = 0b00000001;      //Set prescaler to TMRO 1:4
      TMR0 = 0xd;               //Preset TMRO to overflow on 250 counts
      while(!T0IF);             //Stay until TMRO overflow flag equals 1
      T0IF = 0;                 //Clear the TMR0 overflow flag
}
```

How It Works

After the title block the pic.h file is included to pull in all the necessary compiler information. This is followed by the same configuration setup used in the previous projects.

```
#include <pic.h>          // Include HITECH CC header file

/*
PIC16F690 Configuration
*/
__CONFIG (INTIO & WDTDIS & MCLRDIS & UNPROTECT );
//Internal clock, Watchdog off, MCLR off, Code Unprotected
```

The next section establishes an unsigned short variable called "delay". In the same line the variable is created, an equal sign after it followed by the value of 500 will initialize the variable to hold the decimal value 500.

```
unsigned short delay=500; //Initialize on/off delay value to 500 msec
```

The program will use two functions placed at the bottom of the program listing so we have to let the compiler know they exist prior to entering the main loop. These declarations are called the function prototypes and they are just the first line of the function with a semi-colon added at the end. The *pause* function and the *msecbase* function are declared.

```
void Pause( unsigned short msvalue ); //Establish pause routine
                                      //function
```

```
void msecbase( void );  //Establish millisecond base function
```

The main loop is entered with some register initializations for the PIC16F690 this project will use. The A/D register ANSEL is set to all zeros to make the analog pins all digital mode. The comparator registers CM1CON0 and CM2CON0 are set to all zeros to turn the comparators off. Next the PORTC data register is set to all zeros to preset the I/O to low levels (ground) and finally the TRISC register is set to all zeros to make all the PORTC I/O to outputs. If you combine all this, the PORTC I/O pins will all be digital outputs that start out at a low or ground level, which presets the LEDs to off.

```
main()
{
ANSEL = 0;                      // Intialize A/D ports off
CM1CON0 = 0;                    // Initialize Comparator 1 off
CM2CON0 = 0;                    // Initialize Coparator 2 off

PORTC = 0x00;                   //Clear PortB port
TRISC = 0x00;                   //All PortB I/O outputs
```

Within the main loop is the real functional loop that starts with a *while* statement. The expression 1==1 will always be true so the code within the *while* statements curly brackets will run continuously.

```
while(1==1)                     //loop forever
      {
```

A 16 bit variable "x" is declared prior to entering a *for loop*. In the *for loop* the value of the variable will start at the value one and the loop will continue until the value of x is equal to nine or higher. The value of variable "x" will increment by a factor of two with a simple multiplication, equation of x=x*2. This will make x change from 1 to 2 to 4 and then to 8.

```
      int x;
      for(x=1; x<9; x=x*2)
            {
```

As long as the value of "x" is less than nine the code in *for loop* set of curly brackets will function. Within the curly brackets the data register for PORTC is set equal to the value of variable "x". The first time through "x" is equal to one. The lower four bits of "x" is therefore equal to 0001 binary.

This is how the value will show up on the LED's. A zero will turn that LED off; a one will light the LED. Therefore when x is 0001 binary only the DS1 LED will light up only.

```
            PORTC = x;                // Turn on next LED
```

The next line calls the PAUSE function and passes the variable "delay" to the function. The variable "delay" was preset to 500 earlier so the value of 500 is passed to the PAUSE function. This will create a 500 millisecond delay that I will explain when we get to the PAUSE function.

```
            Pause(delay);             // Delay for .5 seconds
```

The second time through the *for loop*, the value of *x* will equal two and light the second LED DS2 (binary 0010). The third time *x* will equal four or binary 0100 and finally *x* will equal eight in the last loop or 1000 binary. You can see how multiplying by two or doubling the value of *x* makes the lit LED move.

```
            }      // End For
      }     //End while
}     //end main
```

All the curly brackets end together and then the functions are entered. The first is the *pause* function. The *pause* receives the *delay* variable value of 500 and places it in the variable *msvalue* The variable *msvalue* is declared with the function.

```
//*****************************************************
//pause - multiple millisecond delay routine
//*****************************************************

void Pause( unsigned short msvalue )
{
```

Within the function another variable named *x* is created. Now you may wonder how the program can determine which *x* to work with since the program created a variable *x* earlier. The compiler handles this because the *x* that the *pause* function creates is a local variable that only the *pause* function can operate on. After the *pause* function completes its task the RAM space used for the local *x* variable in left available for the next local variable that may be created.

```
unsigned short x;
```

Another *for loop* is used within the function to continuously loop until the value of local *x* is greater than *msvalue* which was set to 500 at the start of the function. I used one of those shortcuts x++ which is the same as x=x+1 just to show how it would be used.

```
for (x=0; x<=msvalue; x++)        //Loop through a delay equal to
      {                           // msvalue in milliseconds.
```

The *pause* function calls the *msecbase* function. Since the value of "msvalue" is 500, the *pause* function is calling the *msecbase* function 500 times.

```
            msecbase();         //Jump to millisec delay routine
            }
}
```

The *msecbase* function uses the TIMER0 of the PIC16F690 to create an accurate one millisecond time base. Since this is called 500 times by the *pause* function, these two functions combine to create a 500 millisecond delay in between updating the LED's. The previous project "How It Works" describes the timer operation so I won't repeat it here.

```
//*******************************************************
//msecbase - 1 msec pause routine
//*******************************************************

void msecbase(void)
{
      OPTION = 0b00000001;      //Set prescaler to TMRO 1:4
      TMR0 = 0xd;               //Preset TMRO to overflow on 250 counts
      while(!T0IF);             //Stay until TMRO overflow flag equals 1
      T0IF = 0;                 //Clear the TMR0 overflow flag
}
```

Next Steps

The value of the "delay" variable can be changed at the top of the program to a larger or smaller value to make the LED's scroll slower or faster respectively. You cold also change the main loop's *for* loop to increment the "x" variable by just one value (i.e. x=x+1). This will change this project from scrolling of the LED's to a binary count display. If you've ever seen one of those binary clocks, this could be used to build one of those.

The for loop can control more LEDs than the four used here with very little modification. If you try this example on a board with eight LEDs all tied to the PORTC pins then you only would need to change the x<9 expression to x<129.

Chapter 12 – Project 5: If –Else

So far all the projects we've completed involve driving an I/O pin as an output. Now let's learn how to initialize a pin as an input and read a switch. A switch press is known as a random event or asynchronous event. In other words, the microcontroller will have to wait around and watch for a switch press. A simple way to do that is to use the *if-else* statement. It will test an input pin for a particular state and *if* that state is sensed then a set of code will be executed. If that state is not sensed then a different set of code will be executed as designated by the *else* statement.

In the case of reading a switch, the circuitry will determine what if a closed switch is a high state or a low state. The software then has to be written so the microcontroller is looking for the proper signal. The PICkit 2 development board has a switch built in and is wired as an active low switch. This just means that when the switch is pressed, the micro pin will see a low level or ground. When the switch is not pressed or at idle, then the micro will see a high level or 5 volts. This project will read the state of the switch and if it is pressed the micro will light the DS1 LED. If the switch is not pressed then the micro will shut the LED off.

Project Setup

The PICkit 2 development board has a normally open switch wired as an active low switch connected to the RA3 pin. The schematic in Figure 12-2 shows the circuitry. You may also notice that the programming header's Vpp pin is also connected to the RA3 pin. This actually causes a conflict when running the PICkit 2 from Microchip's MPLAB like we are using here. There are two ways to get around this. We could just run the development board from a separate power supply and disconnect the PICkit 2 after the chip is programmed but there is an easier choice. The MPLAB software can be setup to release control of that pin automatically after programming the PIC16F690. This is the method I recommend so I show you how to set that up.

PICkit 2 Settings

Under the Programmer>Settings menu option in the MPLAB IDE you can adjust the operation of the PICkit 2. The menu I describe is shown in Figure 12-1. After you click on the Settings option the PICkit 2 settings window shown in Figure 12-2 will appear.

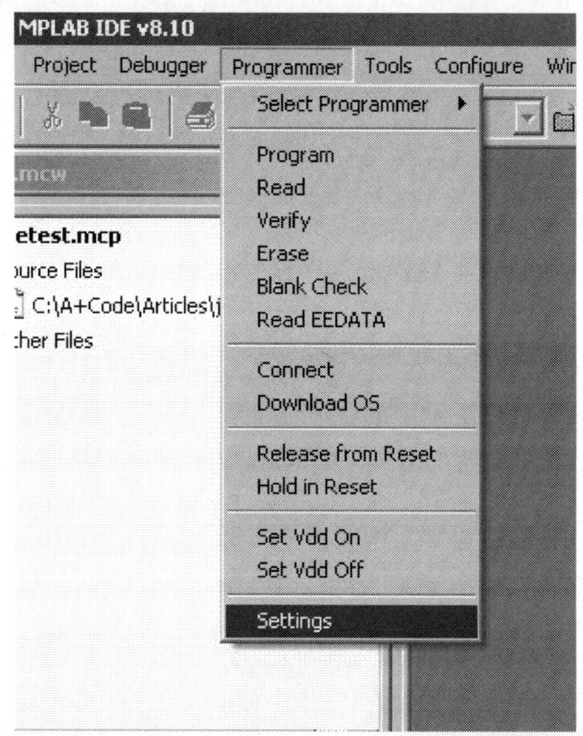

Figure 12-1: MPLAB Programmer Settings Option

The PICkit 2 settings window should have the "Connect at startup" option checked. If it doesn't then I recommend you click in that box. At the bottom you will also see the option to "3-State on "Release from Reset"". Click on the option to enable it and this will disconnect the PICkit 2 from the development board's MCLR pin when the programming is complete. Enabling this option will allow the switch on RA3 pin to be used as just an input switch.

Another option you have is to check the "Program after successful build" option and also the "Run after a successful program" option. If these are enabled the PICkit 2 will automatically program the PIC16F690 and then release the RA3 pin immediately after you get a "Build Succeeded" message in the output window.

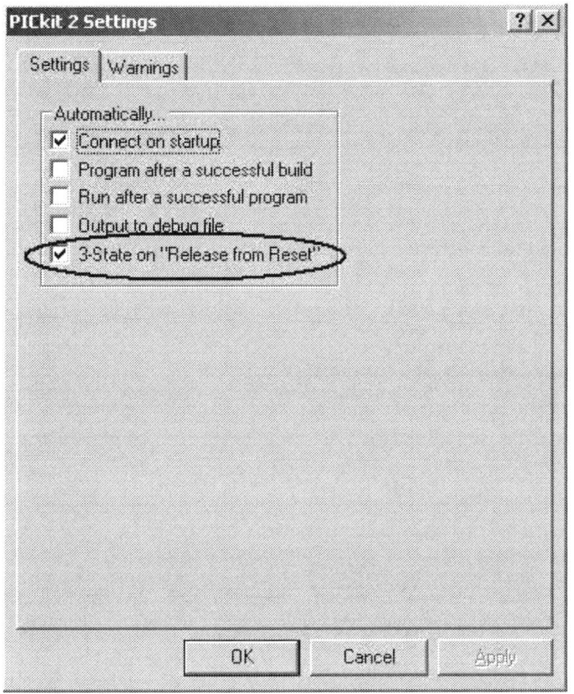

Figure 12-2: PICkit 2 Setting Window

The area of the development board this project will use is shown in Figure 12-3. The switch is labeled SW1. I also show the power connections in the upper left hand corner. I soldered in a pair of header pins so I could connect a battery pack to the board. Using the PICkit 2 settings options allows me to continue powering the development board from the PICkit 2.

Figure 12-3: Project 5 Hardware

The schematic in Figure 12-4 shows the switch configuration. A 10k resistor is used as a pull-up to the Vdd line. Then a 1k resistor is used as a series connection to the RA3 pin. When the switch is not pressed the current will flow from Vdd through the 10k resistor and then through the 1k into the RA3 pin. The RA3 pin will see a high voltage equal to the Vdd level (around 5v in my setup). When the switch is pressed, the current from the 10k resistor is sent to ground through the switch and the RA3 pin will see the voltage across the switch which is essentially zero or ground. The micro will sense this as a low level.

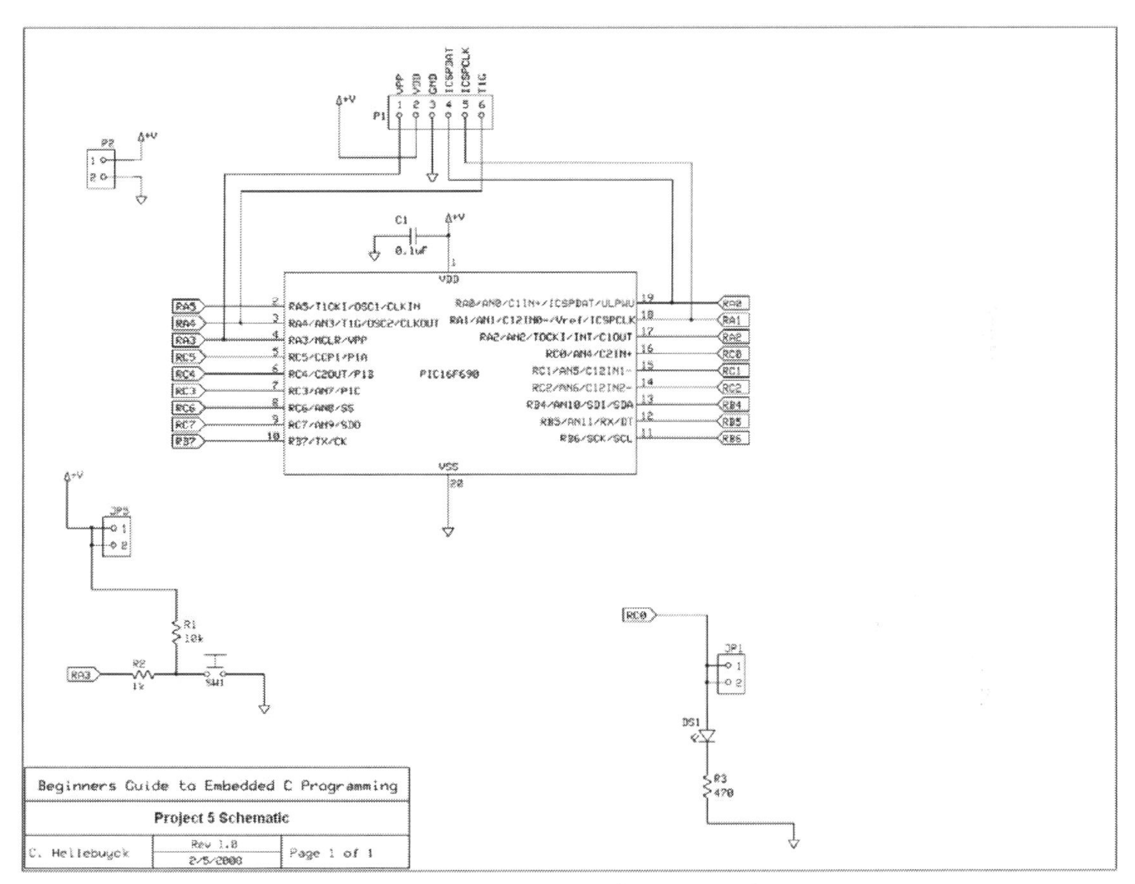

Figure 12-4: Project 5 Schematic

Software

```
/*********************************************************************
*
*
*
*  File:                        Proj5.C
*
*
*
*  Description: This file contains C code for a PIC16F690 using HiTech
*                PICC lite compiler version 9.50 written for the
*                PICkit 2 Demo Board. The project lights the LED on
*                RC1/DS2 but turns it off when the SW1-RA3 switch is
*                pressed. This program uses the If-Else command.
*
*********************************************************************
*
*
*  Created By:  Chuck Hellebuyck 10/19/06
*
*
*
*  Versions:
*
*
*
*  MAJ.MIN - AUTHOR - CHANGE DESCRIPTION
*
*
*
*********************************************************************/

#include <pic.h>        // Include HITECH CC header file

/*
PIC16F690 Configuration
*/
__CONFIG (INTIO & WDTDIS & MCLRDIS & UNPROTECT );
//Internal clock, Watchdog off, MCLR off, Code Unprotected


main()
{
ANSEL = 0;               // Intialize A/D ports off
CM1CON0 = 0;             // Initialize Comparator 1 off
CM2CON0 = 0;             // Initialize Coparator 2 off

PORTC = 0x00;            //Clear PortB port
TRISC = 0x00;            //All PortC I/O outputs
TRISA = 0xFF;            //All PortA I/O inputs

while(1==1)              //loop forever
{
```

```
if (RA3 == 1)              // Test RA3 port
      {
      RC0 = 1;             // If SW1 not pressed, turn on RC0/DS1 LED
      }
else
      {
      RC0 = 0;             // If SW1 pressed, turn off RC1/DS2 LED
      }
}     //End while
}     //end main
```

How It Works

The software starts off as any other project by including the pic.h file and then setting the configuration registers. You can probably do this in your sleep by now. You can also see how you can reuse older programs to build new programs much easier than starting from scratch.

```
#include <pic.h>          // Include HITECH CC header file

/*
PIC16F690 Configuration
*/
__CONFIG (INTIO & WDTDIS & MCLRDIS & UNPROTECT );
//Internal clock, Watchdog off, MCLR off, Code Unprotected
```

The main loop starts off similar as well by setting the registers to make the pins digital and turn the comparators off.

```
main()
{
ANSEL = 0;                // Intialize A/D ports off
CM1CON0 = 0;              // Initialize Comparator 1 off
CM2CON0 = 0;              // Initialize Coparator 2 off
```

The ports are setup once again and this time we add a new line. First the PORTC registers are setup the same as previously because we are driving the LEDs. Then a new register is introduced which is the TRISA register.

```
PORTC = 0x00;             //Clear PortB port
TRISC = 0x00;             //All PortC I/O outputs
```

The TRISA register controls the PORTA pins and we need the RA3 pin to be set to an input. A one in the TRISA register makes the pin an input and a zero makes the pin an output. I was lazy and set all of the PORTA pins to one's. I could of just made RA3 an input by setting PORTA = 0b00001000 which makes all the pins outputs except the 4th bit which represents PORTA pin RA3. This can be a little confusing to the beginner though because they may forget that the port starts off with port pin RA0 so the RA3 pin is actually the 4th bit. I was lazy in the code by not lazy here when I explained it.

```
TRISA = 0xFF;              //All PortA I/O inputs
```

The main loop is entered and a *while* statement creates the main loop.

```
while(1==1)                //loop forever
{
```

The next section includes the *if-else* statement I mentioned at the beginning of this project chapter. The expression being tested is similar to a *while* statement or any other logical expression. The *if* statement will test the RA3 pin and determine if it is at a high level by comparing it against a value of one. If the switch is not pressed then the statement is true and the statements within the first set of brackets are executed. A single line of code between the brackets sets the RC0 pin to a zero or low level that turns the LED to off.

```
if (RA3 == 1)              // Test RA3 port
      {
      RC0 = 0;             // If SW1 not pressed, turn on RC0/DS1 LED
      }
```

If the switch was pressed the RA3 pin will be low and the *if* statement line will not be true. In this case the statements within the *else* brackets are executed which in this example sets the RC0 pin to a high or 1 and this will light the LED.

```
else
      {
      RC0 = 1;             // If SW1 pressed, turn off RC1/DS2 LED
      }
```

The *while* loop will continue this *if-else* test over and over so as the switch changes from pressed to not pressed the LED will light or go out accordingly.

```
}     //End while
}     //end main
```

Next Steps

As usual, there are many ways to modify this simple example. You could change the direction to make the LED light when the switch is idle and turn off the LED when the switch is pressed. You could also add another set of lines to control two LEDs. Light the DS1 LED and turn off the DS2 LED when the switch is pressed and light the DS2 LED and turn off the DS1 LED when the switch is released. This will create a back and forth motion of light.

I mentioned in the text about setting the TRISA = 0b00001000 to just set the RA3 pin. This would be a great time to try that to see if you can make that method work. You could just comment out the original line and add this line to offer both methods for reference.

As a final option I recommend adding another switch to the development board that connects to a different pin. If you connect to the RA2 pin you won't have the Vpp/MCLR interaction. I thought about doing this on this project but decided that some readers may just want to stay with the existing development board or maybe this book is used in a classroom where students are not allowed to modify the board.

Chapter 13 – Project 6: Linking

Linking files together was a new concept to me when I started programming in C. I didn't totally understand it and didn't see the advantage. I have since changed my opinion and knowing how to link files together is fundamental to learning C. The purpose of this project is to demonstrate how to link files and functions together. All the previous projects that used functions had them in the same main.c file but this is not a common practice for the C language programmer. For example, in Chapter 10 I introduced functions by introducing two functions called *pause* and *msecbase*. These two functions formed an accurate time delay that I used between setting LED's on and off. What if I wanted to share the delay routines with another person? Maybe they don't need my same main loop of code but wanted to use the same set of functions for their time delay. They could obviously just cut and paste those sections but that is not always the best method. As programs get bigger it becomes much more difficult to cut and paste.

The advantage to most C compilers including the PICC-Lite compiler is they offer the ability to link many files together into one big project. Each function or common set of functions can be placed in their own file with a .c extension. You could even place all your variables or setup details like configuration and including the pic.h file in one setup header file so you don't have to type it over and over again.

The difference between these two examples is the functions are .c files and the other setup data is put in a .h file. The .c files need to be written so they can be compiled separately where the .h files do not have that requirement. The linker will typically compile each .c file individually and call in the necessary .h files during that process. When all the files have been compiled, the linker then combines the memory requirements of each file into one linked file so a single binary .hex file is created for programming the microcontroller.

This project will just flash the DS1 LED and use the *pause* and *msecbase* functions but the functions will be contained in separate .c files.

Project Setup

The section of the development board used for this project is the same areas we used in the Chapter 9 project that flashed an LED. The schematic is the same as well since this really involves unique software not hardware. Figures 13-1 and 13-2 show the hardware used. It's just the PICkit 2 demo board once again.

Figure 13-1: Project 6 Hardware

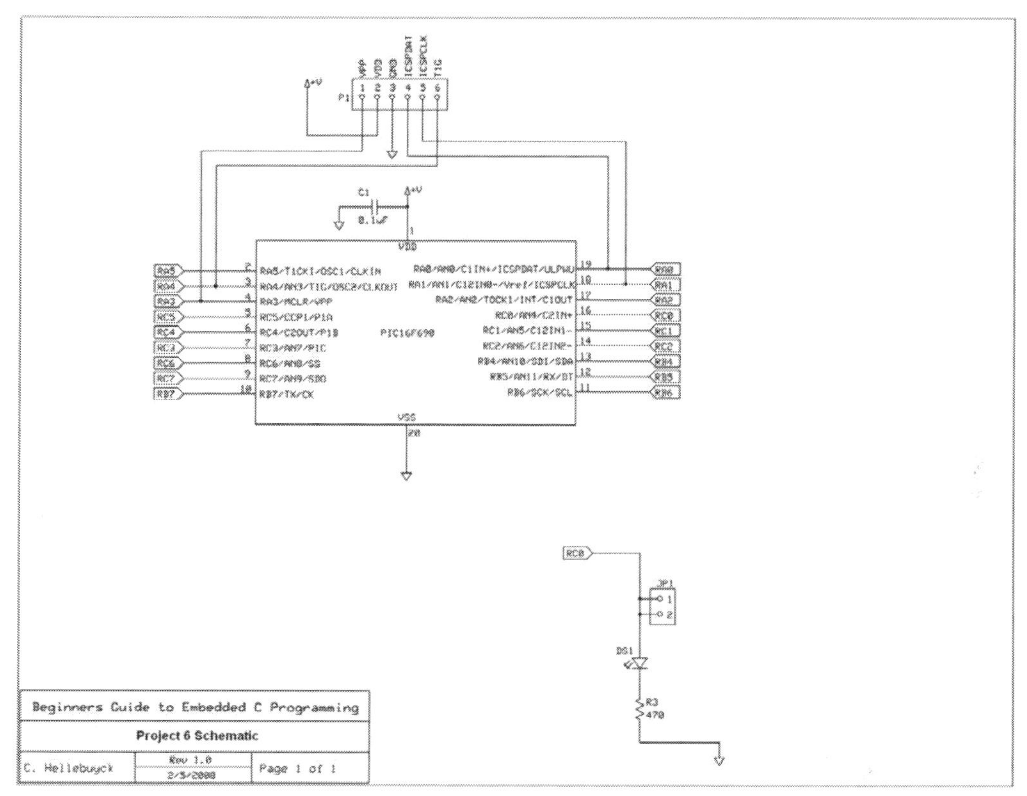

Figure 13-2: Project 6 Schematic

Software

The software is broken into the three files; main.c, pause.c and msecbase.c. These files need to be able to compile separately so there is definitely some redundant command lines. This is the price of linking but this allows these files to easily be shared between users.

MAIN.C

```
/***********************************************************************
*
*  Project 6
*  File:                         main.c
*
*  Description: This file contains C code for a PIC16F690 using HiTech
*                  PICC lite compiler version 9.60 written for the
*                  PICkit 2 Starter Kit.
*                  The project simply flashes the DS1 LED at a 500 msec
*                  rate.
*                  This program demonstrates how to link files or
*                  functions that are in separate files.
```

```
*
*
*
*
*****************************************************************
*
*
*  Created By:  Chuck Hellebuyck 10/18/06
*
*
*  Versions:
*
*
*  MAJ.MIN - AUTHOR - CHANGE DESCRIPTION
*
*
****************************************************************/

#include <pic.h>        // Include HITECH CC header file

/*
PIC16F690 Configuration
*/
__CONFIG (INTIO & WDTDIS & MCLRDIS & UNPROTECT );
//Internal clock, Watchdog off, MCLR off, Code Unprotected


unsigned short delay=500; //Initialize on/off delay value to 500 msec

void pause( unsigned short usvalue );//Establish pause routine function


main()
{
PORTC = 0;                    //Clear PortB port
TRISC = 0;                    //All PortB I/O outputs


while(1==1)                   //loop forever
{
      RC0 = 1;                // Turn on RB0 LED
      pause(delay);           // Delay for .5 seconds
      RC0 = 0;                // Turn off RB0 LED
      pause(delay);           // Delay for .5 seconds


}     //End while
}     //end main
```

PAUSE.C

```
//*****************************************************
//pause - multiple millisecond delay routine
//*****************************************************
#include <pic.h>        // Include HITECH CC header file

void msecbase( void );  //Establish millisecond base function

void pause( unsigned short usvalue )

{
      unsigned short x;

      for (x=0; x<=usvalue; x++)//Loop through a delay equal to usvalue
            {                   // in milliseconds.
            msecbase();         //Jump to millisec delay routine
            }
}
```

MSECBASE.C

```
/*****************************************************
* msecbase - 1 msec pause routine
* The Internal oscillator is set to 4 Mhz and the
* internal instruction clock is 1/4 of the oscillator.
* This makes the internal instruction clock 1 Mhz or
* 1 usec per clock pulse.
* Using the 1:4 prescaler on the clock input to Timer0
* slows the Timer0 count increment to 1 count/4 usec.
* Therefore 250 counts of the Timer0 would make a one
* millisecond delay (250 * 4 usec). But there are other
* instructions in the delay loop so using the MPLAB
* stopwatch we find that we need Timer0 to overflow at
* 243 clock ticks. Preset Timer0 to 13 (0D hex) to make
* Timer0 overflow at 243 clock ticks (256-13 = 243).
* This results in a 1.001 millisecond delay which is
* close enough.
*****************************************************/
#include <pic.h>        // Include HITECH CC header file

void msecbase(void)
{
      OPTION = 0b00000001;    //Set prescaler to TMRO 1:4
      TMR0 = 0xD;             //Preset TMRO to overflow on 250 counts
      while(!T0IF);           //Stay until TMRO overflow flag equals 1
      T0IF = 0;               //Clear the TMR0 overflow flag
}
```

How It Works

The MPLAB software makes this easy to complete. In Figure 13-3 I show the MPLAB window from when I completed the project. The files can each be found in the explorer window at the lower left of the figure. The source files include all three files.

Note:
In the editor window I have the three files tabbed which is an option in MPLAB. If you put your mouse inside one of the open files and right click on your mouse a pop up menu will appear. At the bottom is a "Properties" option. Click on that and another window will appear that has a few boxes you can check. One of the boxes is an option to "Use Tabbed Window". Put a check in that box and click on OK. The next time you open MPLAB the files will be tabbed instead of separate windows.

Each file can now be modified as necessary and when you have all the files the way you want them just click on the build button as usual and the files will be compiled and then linked into a single .hex file. The output window will show that the build succeeded.

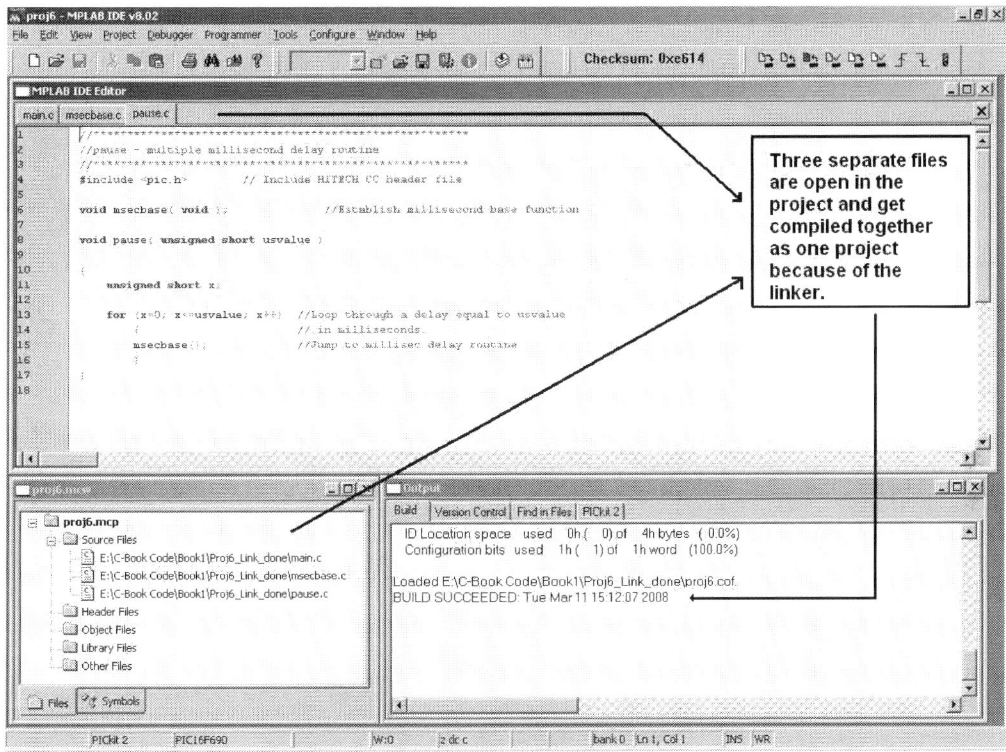

Figure 13-3: MPLAB Window

MAIN.C

The files don't need much explanation if you've done the other projects as they are essentially the same code just broken up into three parts. What you will notice is that each file includes the pic.h file separately. This is required so they can each build separately and independently.

```
#include <pic.h>        // Include HITECH CC header file

/*
PIC16F690 Configuration
*/
__CONFIG (INTIO & WDTDIS & MCLRDIS & UNPROTECT );
//Internal clock, Watchdog off, MCLR off, Code Unprotected
```

As you can see the main.c file has a prototype declared for the *pause* function but not for the *msecbase* function. This is because the pause.c file calls that routine not the main.c. Main.c doesn't even know or care that it exists, it just wants a delay based on the pause.c function.

```
unsigned short delay=500; //Initialize on/off delay value to 500 msec

void pause( unsigned short usvalue );//Establish pause routine function
```

All the port setup and port control of the LED's is in the main.c file. The pause.c and msecbase.c files have nothing to do with driving the outputs so this section stays with the main.c file.

```
main()
{
PORTC = 0;                    //Clear PortB port
TRISC = 0;                    //All PortB I/O outputs


while(1==1)                   //loop forever
{
      RC0 = 1;                // Turn on RB0 LED
      pause(delay);           // Delay for .5 seconds
      RC0 = 0;                // Turn off RB0 LED
      pause(delay);           // Delay for .5 seconds



}     //End while
}     //end main
```

The pause.c file has a prototype declared for the *msecbase* function that is called in its main loop of code. The prototype was above the main.c file when these were all in one single file. In order for pause.c to compile alone it needs the pic.h file included and the *msecbase* prototype declared.

PAUSE.C

```
//*****************************************************
//pause - multiple millisecond delay routine
//*****************************************************
#include <pic.h>          // Include HITECH CC header file

void msecbase( void );  //Establish millisecond base function

void pause( unsigned short usvalue )

{
```

```
      unsigned short x;

      for (x=0; x<=usvalue; x++)//Loop through a delay equal to usvalue
            {                    // in milliseconds.
            msecbase();          //Jump to millisec delay routine
            }
}
```

The msecbase.c file also calls the pic.h file so it can compile successfully. Everything else in this file is the same as when it was part of the big file. You can see it didn't take much to make these into separate files but I hope you begin to see that breaking them up does offer some advantages to keeping track of what is going on in each section of code.

MSECBASE.C

```
#include <pic.h>        // Include HITECH CC header file

void msecbase(void)
{
      OPTION = 0b00000001;     //Set prescaler to TMRO 1:4
      TMR0 = 0xD;              //Preset TMRO to overflow on 250 counts
      while(!T0IF);            //Stay until TMRO overflow flag equals 1
      T0IF = 0;                //Clear the TMR0 overflow flag
}
```

Next Steps

There are so many next steps you could take here. You could create one main.c that flashes the DS1 LED and then another that flashes the DS2 LED. They don't have to be called main.c either they just need to have a main loop. One could be called DS1.c and the other DS2.c. Now you can link in which file you want based on which LED you want to flash. You could also try to create a faster time base function and have it inside a separate .c file such as microsecbase.c. With this option the pause.c file could be modified to call a slow time base or a faster time base with very little modification. You would just link in the time base file you need. By just changing the time base function call you will have a shorter delay for doing things like producing a fast square wave that doesn't need to run slow enough for the human eye to see like an LED requires.

This is a very important topic for the beginner to learn so I hope this example explained linking well enough for you to be comfortable with it. I'm going to use it again in the future projects to help drive home the technique.

Chapter 14 – Project 7: Switch-Case

The *switch-case* statement is a very popular statement for creating a lookup table structure which I will cover in this project. I will also expand on the linking project in the previous chapter by showing how to include a custom header file (.h file). This project will read the switch on the PICkit 2 development board and increment the value of a variable based on how many times the switch was pressed. The *switch-case* statement will then light LED's based on the value of the variable in a look-up table type of arrangement.

Project Setup

The project will use all four LEDs and the switch. Because the switch is tied to the MCLR pin of the PIC16F690 and that pin is also connected to the PICkit 2 reset line, you should make sure the PICkit 2 settings still have the "3-State" option selected. Once it's set it should not change but its best to verify it so you are not trying to debug something that has nothing to do with your code. Figure 14-1 shows the areas of the development board again that will be used in this project.

Figure 14-1: Project 7 Hardware

The schematic in Figure 14-2 shows the connections being used. If you decided to use a separate power supply, the PICkit 2 will actually sense if there is power on the development board and supply it or not supply it from the USB port so you can actually run all the projects from the separate battery approach if you desire.

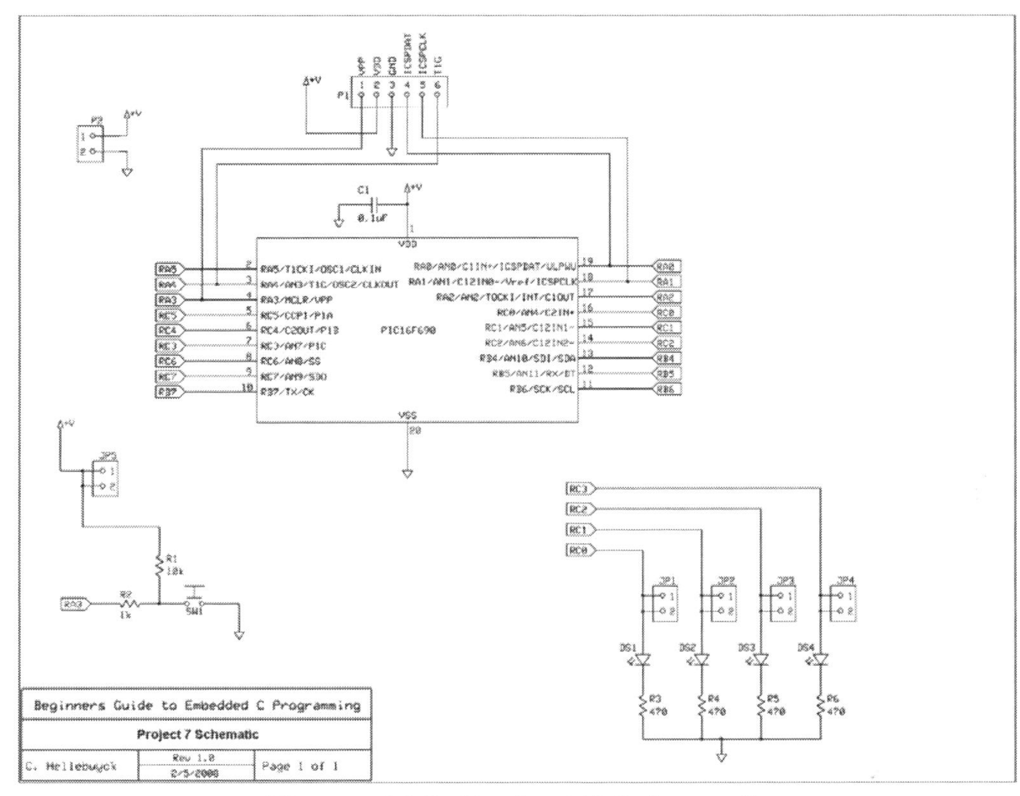

Figure 14-2: Project 7 Schematic

Software

The software is composed of multiple files that get linked together. The "proj7.c" file is the main program that calls in the function: "pause.c". To use the "pause.c" file though, a prototype must be declared and that prototype is contained in the "pause.h" header file included at the top of the "proj7.c" file. Notice that the main loop is contained in a file that isn't called main.c. It's not a requirement to have a main.c file but you need a file that contains the main loop. I like calling the main file main.c because as the number of files being linked together gets larger, it's easier to find out where it all starts by just looking for the main.c file.

Within "pause.c" a header file will also get included called "msecbase.h". It contains the prototype for the *msecbase* function that the "pause.c" file requires.

The “msecbase.c” file is the final file needed but doesn’t require anything additional so no header file is required. All these files get grouped together by the built in linker. The linker is a separate application that is built into the PICC-Lite compiler. Some C compilers will make this a completely separate file. I like it hidden as the PICC-Lite compiler does. The How it Works section will explain each of these files in more detail.

```
/*********************************************************************
*
*
*  File:                          Proj7.c
*
*  Description: This file contains C code for a PIC16F690 using HiTech
*                  PICC lite compiler version 9.60 written for the
*                  PICkit 2 Demo Board. The project lights the LEDs in
*                  sequence as the switch SW1-RA3 switch is pressed.
*                  This program uses the Switch-Case command to control
*                  which LEDs to light.
*
*                 This program also uses linking to include a debounce
*                  delay.
**********************************************************************
*
*
*  Created By:  Chuck Hellebuyck 7/12/07
*
*
*  Versions:
*
*
*  MAJ.MIN - AUTHOR - CHANGE DESCRIPTION
*
*
*
*******************************************************************/

#include <pic.h>                // Include HITECH CC header file
#include "pause.h"              // Include Pause Function Header

#define SW1        RA3          // Create nickname for RA3 port

/*
PIC16F690 Configuration
*/
__CONFIG (INTIO & WDTDIS & MCLRDIS & UNPROTECT );
//Internal clock, Watchdog off, MCLR off, Code Unprotected
```

```
main()
{
unsigned char state_led = 0;  //Create 8 bit variable to store switch
                              // count

ANSEL = 0;                    // Intialize A/D ports off
CM1CON0 = 0;                  // Initialize Comparator 1 off
CM2CON0 = 0;                  // Initialize Coparator 2 off

PORTC = 0x00;                 //Clear PortB port
TRISC = 0x00;                 //All PortC I/O outputs
TRISA = 0xFF;                 //All PortA I/O inputs


while(1==1)                   //loop forever
{

if (SW1 == 0)                 // if a button press is detected,
      {
      state_led = state_led+1; //  increment the LED state variable
      switch (state_led)
             {
                   case 1:          // STATE0: turn only the D0 LED on
                         RC0 = 1;
                         RC1 = 0;
                         RC2 = 0;
                         RC3 = 0;
                         break;
                   case 2:          // STATE1: turn only the D1 LED on
                         RC0 = 0;
                         RC1 = 1;
                         RC2 = 0;
                         RC3 = 0;
                         break;
                   case 3:          // STATE2: turn only the D2 LED on
                         RC0 = 0;
                         RC1 = 0;
                         RC2 = 1;
                         RC3 = 0;
                         break;
                   case 4:          // STATE3: turn only the D3 LED on
                         RC0 = 0;
                         RC1 = 0;
                         RC2 = 0;
                         RC3 = 1;
                         break;
                   default:         // If state_led > 3 reset switch
                                    //  count to zero
                         state_led = 0; // All LEDs off
                         RC0 = 0;
                         RC1 = 0;
                         RC2 = 0;
                         RC3 = 0;
                         break;
```

```
            } //end switch

         while (!SW1);      // Hold here until switch is released
         pause (10);        // Delay 10 milliseconds and check again
         while (!SW1);      // For simple debounce of switch
      } //End If
}     //End while
}     //end main
```

PAUSE.H

```
/* Header file for Pause.C */

void pause( unsigned short usvalue );//Establish pause routine function
```

PAUSE.C

```
//*****************************************************
//pause - multiple millisecond delay routine
//*****************************************************
#include <pic.h>         // Include HITECH CC header file
#include "msecbase.h"    // Include msecbase function header file


void pause( unsigned short usvalue )

{
      unsigned short x;

      for (x=0; x<=usvalue; x++)     //Loop through a delay equal to
                                     // usvalue
            {                        // in milliseconds.
            msecbase();              //Jump to millisec delay routine
            }
}
```

MSECBASE.H

```
/* Header file for msecbase.c */

void msecbase( void );        //Establish millisecond base function
```

MSECBASE.C

```
/*****************************************************
* msecbase - 1 msec pause routine
* The Internal oscillator is set to 4 Mhz and the
* internal instruction clock is 1/4 of the oscillator.
* This makes the internal instruction clock 1 Mhz or
* 1 usec per clock pulse.
```

```
* Using the 1:4 prescaler on the clock input to Timer0
* slows the Timer0 count increment to 1 count/4 usec.
* Therefore 250 counts of the Timer0 would make a one
* millisecond delay (250 * 4 usec). But there are other
* instructions in the delay loop so using the MPLAB
* stopwatch we find that we need Timer0 to overflow at
* 243 clock ticks. Preset Timer0 to 13 (0D hex) to make
* Timer0 overflow at 243 clock ticks (256-13 = 243).
* This results in a 1.001 millisecond delay which is
* close enough.
*****************************************************/
#include <pic.h>        // Include HITECH CC header file

void msecbase(void)
{
      OPTION = 0b00000001;      //Set prescaler to TMRO 1:4
      TMR0 = 0xD;               //Preset TMRO to overflow on 250 counts
      while(!T0IF);             //Stay until TMRO overflow flag equals 1
      T0IF = 0;                 //Clear the TMR0 overflow flag
}
```

How It Works

The proj7.c file will be the first file I'll discuss in detail. At the top of this file are two include statements. One is the now common pic.h header file required for the PICC-Lite compiler. The second is the pause.h file that has the prototype statement for the pause function in the file pause.c. This file proj7.c will call the function *pause* so the prototype has to be declared before the function is called.

```
#include <pic.h>              // Include HITECH CC header file
#include "pause.h"            // Include Pause Function Header
```

There is a new statement added that hasn't been used before and that is a *#define* statement. This is used to establish a nickname for a constant value. That constant value is port pin RA3 which is the PICC-Lite name for the I/O pin PORTA pin 3. This *#define* statement creates the nickname "SW1" for the RA3 pin. After this line the compiler will replace any SW1 with RA3 prior to compiling. It's easier to understand the code when the label SW1 is used because that directly correlates to the SW1 wording on the development board.

Note:
SW1 is capitalized. Constants are typically capitals or at least the first letter is so you can easily distinguish between constants and variables. This is a C language recommended practice.

```
#define SW1       RA3         // Create nickname for RA3 port
```

The configuration bits are set once again to the same setup as usual.

```
/*
PIC16F690 Configuration
*/
__CONFIG (INTIO & WDTDIS & MCLRDIS & UNPROTECT );
//Internal clock, Watchdog off, MCLR off, Code Unprotected
```

The main loop of code starts off by creating an eight bit wide variable called "state_led" and presets it to zero. This will store the number of times that the SW1 switch is pressed.

```
main()
{
unsigned char state_led = 0;  //Create 8 bit variable to store switch
                              // count
```

A series of special function registers are initialized to make all pins digital with PORTA all inputs and PORTC all outputs.

```
ANSEL = 0;                    // Intialize A/D ports off
CM1CON0 = 0;                  // Initialize Comparator 1 off
CM2CON0 = 0;                  // Initialize Coparator 2 off

PORTC = 0x00;                 //Clear PortC port
TRISC = 0x00;                 //All PortC I/O outputs
TRISA = 0xFF;                 //All PortA I/O inputs
```

The central *while* loop, which is continuous, is entered after this. The first operation in the loop is to test the switch to see if it is pressed. The hardware is wired to make the SW1 input a zero or ground when the switch is pressed. The next line is an *if* statement that checks the state of the SW1 input. If the switch is not pressed then everything within the *if* statement curly brackets is by passed which will put the program execution back up to the *while* loop which will test the switch state again. If the input is low then the switch was pressed and the "state_led" variable will be incremented by one.

```
while(1==1)                    //loop forever
{

if (SW1 == 0)                  // if a button press is detected,
      {
      state_led = state_led+1; //  increment the LED state variable
```

After the "state_led" variable is incremented, the next command line is the *switch-case* statement. The *switch-case* structure will light the LEDs on the development board according to the value in the "state_led" variable. Each potential value of the "state-led" variable is specified as a separate case. The first one is "case 1:". You can think of this as "in the case when state_led equals one do the following". The lines that follow the "case 1:" implement the desired LED lighting. The DS1 LED is connected to the RC0 I/O pin and turned on while all the other LEDs are turned off. The last command is a BREAK statement which is required to end the case 1 actions. The *break* statement jumps the program control to outside the *switch-case* structure which ends at the closing curly bracket as shown by the comment line "//end switch".

```
      switch (state_led)
             {
                  case 1:            // STATE0: turn only the D0 LED on
                        RC0 = 1;
                        RC1 = 0;
                        RC2 = 0;
                        RC3 = 0;
                        break;
```

The *switch-case* statement continues to list other options for when "state_led" is equal to two, three or four.

```
                  case 2:            // STATE1: turn only the D1 LED on
                        RC0 = 0;
                        RC1 = 1;
                        RC2 = 0;
                        RC3 = 0;
                        break;
                  case 3:            // STATE2: turn only the D2 LED on
                        RC0 = 0;
                        RC1 = 0;
                        RC2 = 1;
                        RC3 = 0;
                        break;
                  case 4:            // STATE3: turn only the D3 LED on
                        RC0 = 0;
```

```
        RC1 = 0;
        RC2 = 0;
        RC3 = 1;
        break;
```

The last case in the list of options is the *default* choice. When none of the *case* values match the value of the the *switch* statement variable, the commands under the *default* choice are implemented. This can be used as an error state if your program has a *case* line for every possible choice you expect. If for some strange reason the variable gets corrupted and has a value that doesn't match any of the *case* statements then the *default* becomes the error state.

In this example I use the *default* as a reset mechanism rather than an error state. I fully expect the value of "state_led" to count up to 5 which has no *case* statement to match. Therefore the *default* is executed. In the *default* code, the value of "state_led" is reset to zero and the LEDs are all turned off. The *break* statement jumps the program out to the statement after the "//end switch" line.

```
    default:            // If state_led > 3 reset switch
                        //  count to zero
        state_led = 0;  // All LEDs off
        RC0 = 0;
        RC1 = 0;
        RC2 = 0;
        RC3 = 0;
        break;
    } //end switch
```

The statements that follow the *switch-case* statement act as a debounce routine for the switch. When a mechanical switch is pressed or released, the internal spring metal will fluctuate like a car spring with a worn out shock absorber. This is known as switch bounce and can create an electrical signal that bounces between connected and disconnected quickly. The microcontroller runs much faster than the switch bounce and it can read that switch bounce as sequential open and closed switch connections.

When the switch has been pressed the program entered the *if* statement routine and the *switch-case* statement went into action. When the *switch-case* is complete, the switch may still be bouncing or our finger may still be actually pressing it. We don't want the routine to loop back to the top until we've seen the switch released because a single press of the switch could end up looking like multiple presses. The routine holds up the program until the switch is released and it does this in a simple way.

A *while* statement is used to test the switch input for the switch to be released. We know that on this hardware setup a switch press will look like a zero or ground voltage level therefore we wait via the *while* statement for the switch input SW1 to be logically true or a one before moving on. This is opposite to the way we've used the *while* statement previously. We want to stay at the *while* statement when SW1 is low or false. Therefore we invert the logic with an exclamation point. It's the same as saying wait until SW1 is not true or zero. When the switch is released, the SW1 signal should be high or 1 so it's true therefore the *while* statement releases program control to the next command line.

```
            while (!SW1);      // Hold here until switch is released
```

That high signal that the micro sensed could be a switch that was released completely or it could be a high signal in the middle of the debounce noise. Therefore we need to give it a little time to settle out before we test it again to verify the switch really is released. A simple call to the *pause* function with a value of 10 will create a 10 millisecond delay. This delay is just a guess based on some testing on the development board but I found 10 milliseconds was enough time to let the switch settle out.

```
            pause (10);        // Delay 10 milliseconds and check again
```

The switch input is tested again to see if it is still high with a *while* statement and then the program will jump back to the top because it is still within the while (1==1) loop.

```
            while (!SW1);      // For simple debounce of switch
      } //End If
}     //End while
}     //end main
```

The pause.h file simply contains the *pause* function prototype line. The linker will just insert the contents of this file into the proj7.c file at the exact location where the pause.h file was included. If you had a lot of functions already created then using a separate header file for each function prototype makes it easier to just include or not include all the required header files in the main program.

PAUSE.H

```
/* Header file for Pause.C */

void pause( unsigned short usvalue );//Establish pause routine function
```

The pause.c file is the same *pause* function we used in the earlier projects; it's just separated into its own file. The linker will combine this with the proj7.c file before the compiler generates the binary .hex file.

PAUSE.C

```
//****************************************************
//pause - multiple millisecond delay routine
//****************************************************
```

The *include*s at the top of the function are really the only thing unique when compared to the previous *pause* functions used. Each ".c" file has to compile on its own so the pic.h header file must be included along with any function prototypes being used. In this case the msecbase.h file has the necessary prototype line for the *msecbase* function.

```
#include <pic.h>          // Include HITECH CC header file
#include "msecbase.h"     // Include msecbase function header file
```

The rest of the pause.c file is the same *pause* function I described earlier.

```
void pause( unsigned short usvalue )

{
      unsigned short x;

      for (x=0; x<=usvalue; x++)      //Loop through a delay equal to
                                      // usvalue
            {                         // in milliseconds.
            msecbase();               //Jump to millisec delay routine
            }
}
```

As you can see, the msecbase.h file continues the same technique of creating the inserted function prototype line. This is a simple example but if you wanted to share this *msecbase* function with somebody else, you have made it very simple for them to just include the msecbase.h file at the top of their main program and then just call the *msecbase* function.

MSECBASE.H

```
/* Header file for msecbase.c */

void msecbase( void );  //Establish millisecond base function
```

The msecbase.c file doesn't require a prototype declared since it doesn't call any functions. What it does need though is the pic.h file so it can compile as a standalone file. Other than the pic.h *include* the msecbase.c file is the same setup we've used.

MSECBASE.C

```
#include <pic.h>          // Include HITECH CC header file

void msecbase(void)
{
      OPTION = 0b00000001;     //Set prescaler to TMRO 1:4
      TMR0 = 0xD;              //Preset TMRO to overflow on 250 counts
      while(!T0IF);            //Stay until TMRO overflow flag equals 1
      T0IF = 0;                //Clear the TMR0 overflow flag
}
```

Next Steps

Where to go from here is really an interesting choice. You could try to break up the proj7.c file into more files just to make sure you understand this topic but more importantly was to gain an understanding of the *switch-case* statement. I would recommend you add more combinations of LED lighting by adding more *case* elements. Just insert them between the *case* 4: and the *default*. You can add a *case* 5: and a *case* 6: to see if you can light the LEDs is a 5th and 6th unique pattern. That should be an easy thing to do and will help you understand the *switch-case* operation.

Chapter 15 – Project 8: Reading A/D port

The final project I wanted to cover is how to read a variable voltage with an Analog to Digital (A/D) converter. This will also require the program to setup some of the PIC16F690 registers differently. I'm not going to use multiple files for this project to keep things simpler.

The project will use the potentiometer on the development board as the control. When the potentiometer is turned in one direction, the LEDs will light up in sequence. When it is turned the opposite direction, the LEDs will go out in sequence. It will look very similar to the display you might see on a stereo volume control or the graphic display on a car radio. We won't be changing the volume of anything but we will be creating the visual affect.

Project Setup

The development board makes this easy to setup and the areas of the board that will be used are circled again as shown in Figure 15-1. The LEDs and potentiometer are circled this time. The potentiometer is wired to voltage already so turning the thumbwheel will change the voltage going into the PIC16F690.

Figure 15-1: Project 8 Hardware

The schematic in Figure 15-2 shows the connections to the PIC16F690. The potentiometer is connected to the RA0 pin. The RA0 pin will have to be changed from a digital pin to an analog pin in software. The LEDs will use the same RC0 through RC3 digital I/O pins. Again, software will have to take care of that setup as well. The development board can be powered from the PICkit 2 but you can use an external power source if you desire. Just don't exceed five volts to be safe.

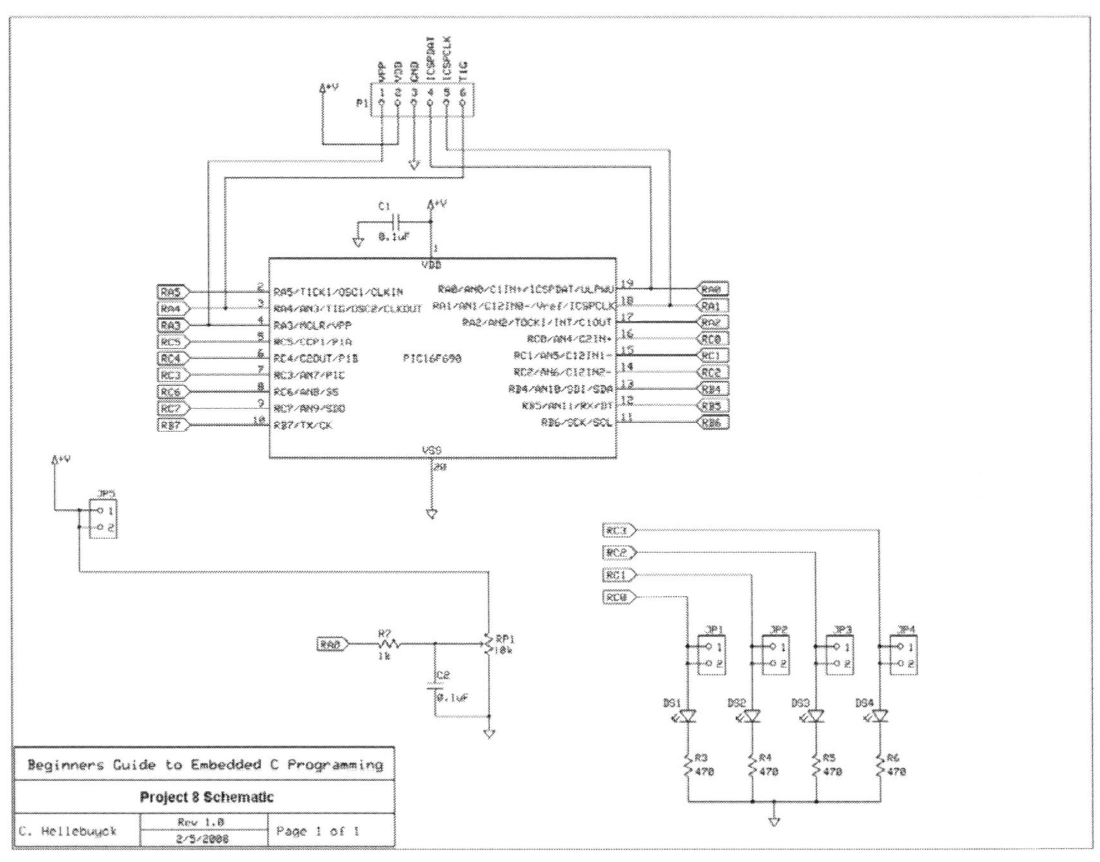

Figure 15-2: Project 8 Schematic

Software

The software listing for project eight is listed below. This is a single file that doesn't use any function calls or have any outside files to link in. I kept this one simple because I really wanted to demonstrate how to access a common peripheral on the PIC16F690 which is the Analog to Digital (A/D) port. The "How It Works" section will explain this peripheral in more detail. If you want to read a sensor or do any kind of measurement of analog signals with an embedded microcontroller, then knowing how to use the A/D port is critical.

```
/**********************************************************************
*
*
*  File:                         Proj8.c
*
*  Project:                      Project 8
*
*
*
*  Description: This file contains C code for a PIC16F690 using HiTech
*                PICC lite compiler version 9.60 written for the P
*                PICkit 2 Demo Board. The project lights the LEDs in
*                sequence as the potentiometer is turned. This program
*                uses the PIC16F690 A/D port RA0 to read the
*                potentiometer value and then compares the value using
*                a series of IF commands to test the value against
*                preset constants to determine which LEDs to light.
*
***********************************************************************
*
*
*  Created By:  Chuck Hellebuyck 7/12/07
*
*
*
*  Versions:
*
*
*
*  MAJ.MIN - AUTHOR - CHANGE DESCRIPTION
*
*
*
**********************************************************************/

#include <pic.h>

/*
PIC16F690 Configuration
```

```
*/
__CONFIG (INTIO & WDTDIS & MCLRDIS & UNPROTECT );
//Internal clock, Watchdog off, MCLR off, Code Unprotected


#define DS1_on 0b00000001;    //Create nickname to light DS1 LED
#define DS2_on 0b00000010;    //Create nickname to light DS2 LED
#define DS3_on 0b00000100;    //Create nickname to light DS3 LED
#define DS4_on 0b00001000;    //Create nickname to light DS4 LED

int advalue = 0;            //Create A/D storage value and clear it

main()
{
PORTA = 0;                  //Clear PortA
TRISA = 0xFF;               //All PortA I/O inputs
CM1CON0 = 0;                //C1 Comparator off
ANSEL = 1;                  //A/D module on
ADCON0 = 0b00000001;//AtoD on, Left justified, Channel AN0/RA0 selected
ADCON1 = 0b00111000;        //Internal RC clock for A/D conversion
PORTC = 0;                  //Clear PortC
TRISC = 0;                  //All PortC is output

while(1==1)                 //Loop Forever
{
GODONE = 1; //Start A/D process
      while (GODONE ==1) //wait for A/D to finish
      {
      }

advalue = ADRESH;           //Store A/D result in variable ADVALUE

      if (advalue <60)      //Test if A/D value is less than 60 decimal
      {
      PORTC = DS4_on;       //Less than 60, light DS4 LED
      }
      if (advalue <120)     //Test if A/D value less than 120 decimal
      {
      PORTC = DS3_on;       //Less than 120, light DS3 LED
      }
      if (advalue <180)     //Test if A/D value less than 180
      {
      PORTC = DS2_on;       //Less than 180, light DS2 LED
      }
      if (advalue < 240) //Test if A/D value less than 240
      {
      PORTC = DS1_on;       //Less than 240, light DS1 LED
      }
      if (advalue >= 240)//Test if A/D value equal to
                          // or greater than 240
      {
      PORTC = 0;            //Set all LEDs off
      }
}     //End While
```

```
}       //End Main
```

How It Works

The program will read the voltage on the potentiometer with an Analog to Digital (A/D) port and then light the LEDs according the value that was read. Seems simple enough but there are several unique registers to setup for this to work. The program starts off in the usual way by including the pic.h file and then establishes the same configuration settings.

```
#include <pic.h>

/*
PIC16F690 Configuration
*/
__CONFIG (INTIO & WDTDIS & MCLRDIS & UNPROTECT );
//Internal clock, Watchdog off, MCLR off, Code Unprotected
```

The next set of command lines use the *#defines* to create some constants that are associated with a specific binary value. These binary values will be used for setting the LEDs to the proper on/off arrangement. Having to type in those binary values over and over is a pain so it's easier to first create nicknames or constant labels for these values. Each constant lights only one LED so the nicknames are DS1_on, DS2_on, etc. This makes it easier to understand which LED is being lit later in the program as well.

```
#define DS1_on 0b00000001;     //Create nickname to light DS1 LED
#define DS2_on 0b00000010;     //Create nickname to light DS2 LED
#define DS3_on 0b00000100;     //Create nickname to light DS3 LED
#define DS4_on 0b00001000;     //Create nickname to light DS4 LED
```

The A/D port on the PIC16F690 will read the voltage of the potentiometer and convert it to a digital value. The PIC16F690 can create an 8-bit or 10-bit value meaning the converted digital value can range from 0 to 256 or 0 to 1024. In this case we don't need a lot of resolution so the 8-bit 0 to 256 range works fine. That digital value will have to be stored somewhere so a variable needs to be created. Since we are using an 8-bit conversion we only need to establish an 8-bit variable but out of habit a 16-bit *int* variable called "advalue" was created and initialized it to zero.

```
int advalue = 0;        //Create A/D storage value and clear it
```

The main loop of code is entered next and a few common register setup lines are established. PORTA data register is cleared and the TRISA register sets all PORTA pins to inputs.

```
main()
{
PORTA = 0;                  //Clear PortA
TRISA = 0xFF;               //All PortA I/O inputs
```

Then the comparators are turned off with the CM1CON0 register set to zero.

```
CM1CON0 = 0;                //C1 Comparator off
```

The next line is dedicated to setting up the PIC16F690's internal A/D converter. The ANSEL register is shown below in Figure 15-3.

REGISTER 4-3: ANSEL: ANALOG SELECT REGISTER

R/W-1	R/W-1	R/W-1	R/W-1	R/W-1	R/W-1	R/W-1	R/W-1
ANS7	ANS6	ANS5	ANS4	ANS3	ANS2	ANS1	ANS0
bit 7							bit 0

Legend:			
R = Readable bit	W = Writable bit	U = Unimplemented bit, read as '0'	
-n = Value at POR	'1' = Bit is set	'0' = Bit is cleared	x = Bit is unknown

bit 7-0 **ANS<7:0>**: Analog Select bits
Analog select between analog or digital function on pins AN<7:0>, respectively.
1 = Analog input. Pin is assigned as analog input[1].
0 = Digital I/O. Pin is assigned to port or special function.

Note 1: Setting a pin to an analog input automatically disables the digital input circuitry, weak pull-ups and interrupt-on-change if available. The corresponding TRIS bit must be set to Input mode in order to allow external control of the voltage on the pin.

Figure 15-3: ANSEL Register

The ANSEL register is short for Analog Select Register and it determines which input pins are connected to the A/D converter and which pins are connected to the digital circuitry. To connect the AN0 pin to the A/D converter we set bit 0 in the ANSEL register. The rest of the bits are set to zero so they are digital inputs. The statement below handles that.

```
ANSEL = 1;                  //A/D module on
```

There are additional registers that setup the A/D converter for the operation we want. Since the PIC16F690 A/D can produce a 10-bit value, the result of the A/D conversion is first stored in two 8-bit wide registers; ADRESH and ADRESL. One of these registers will have all eight bits used and the other will only use two bits used for a total of ten. This means there are six bits that are not used. The result can be shifted to put the eight most significant bits in the ADRESH register and the lower two bits in the ADRESL. This is the best way to handle 8-bit operation because all the bits we need will be in the ADRESH register. Figure 15-4 shows the register setup as I describe. If you read through the PIC16F690 data sheet on the A/D register you will see a better explanation of how the whole A/D converter works. I'm just going through the basics here to allow you to learn the code.

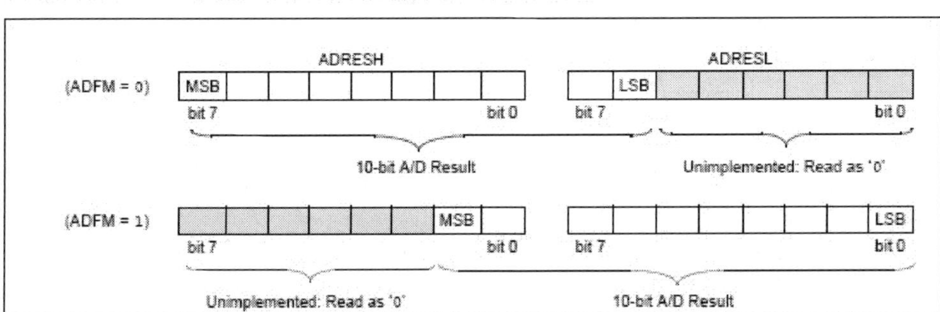

Figure 15-4: A/D Conversion Layout

The ADCON0 register is the A/D control register that determines the left or right shift of the result. The ADCON0 register is shown below in Figure 15-5.

REGISTER 9-1: ADCON0: A/D CONTROL REGISTER 0

R/W-0	R/W-0	R/W-0	R/W-0	R/W-0	R/W-0	R/W-0	R/W-0
ADFM	VCFG	CHS3	CHS2	CHS1	CHS0	GO/$\overline{DONE}$	ADON
bit 7							bit 0

Legend:			
R = Readable bit	W = Writable bit	U = Unimplemented bit, read as '0'	
-n = Value at POR	'1' = Bit is set	'0' = Bit is cleared	x = Bit is unknown

bit 7 **ADFM:** A/D Conversion Result Format Select bit
1 = Right justified
0 = Left justified

bit 6 **VCFG:** Voltage Reference bit
1 = VREF pin
0 = VDD

bit 5-2 **CHS<3:0>: Analog Channel Select bits**
0000 = AN0
0001 = AN1
0010 = AN2
0011 = AN3
0100 = AN4
0101 = AN5
0110 = AN6
0111 = AN7
1000 = AN8
1001 = AN9
1010 = AN10
1011 = AN11
1100 = CVREF
1101 = 0.6V Reference
1110 = Reserved. Do not use.
1111 = Reserved. Do not use.

bit 1 **GO/$\overline{DONE}$:** A/D Conversion Status bit
1 = A/D conversion cycle in progress. Setting this bit starts an A/D conversion cycle.
This bit is automatically cleared by hardware when the A/D conversion has completed.
0 = A/D conversion completed/not in progress

bit 0 **ADON:** ADC Enable bit
1 = ADC is enabled
0 = ADC is disabled and consumes no operating current

Figure 15-5: ADCON0 Register

The ADCON0 also controls the voltage reference source for the A/D. We want Vdd as the A/D reference which is the same voltage driving the PIC16F690. We need to use the AN0 port because the potentiometer is connected to that pin. We previously set that pin to input with the TRISA setting and made it and A/D connection with the ANSEL setting but now we need to tell the A/D converter which pin to convert. Seems like a lot of settings for one pin but that is the way the PIC16F690 internal setup is controlled.

Bits 2 through 5 of the ADCON register selects the A/D pin to convert. Bit1 of the ADCON register is an indicator that changes from 0 to 1 when the A/D conversion is complete. Notice the name of this bit is shown in Figure 15-5 as Go/Done. We will use that name later so remember where this is at. Bit0 of the ADCON enables and disables the whole A/D converter. All this is setup in the command line below. Notice that the Go/Done bit is set to zero.

```
ADCON0 = 0b00000001;//AtoD on, Left justified, Channel AN0/RA0 selected
```

Whew! We're still not done. A second A/D control register needs to be setup. The ADCON1 register selects the clock source for the A/D converter to digitize the sample using a sample and hold circuit. In the setup below we use the internal Frc option. If you need really accurate sampling then you might use the internal system clock.

```
ADCON1 = 0b00111000;     //Internal RC clock for A/D conversion
```

The ADCON1 register is shown in Figure 15-6.

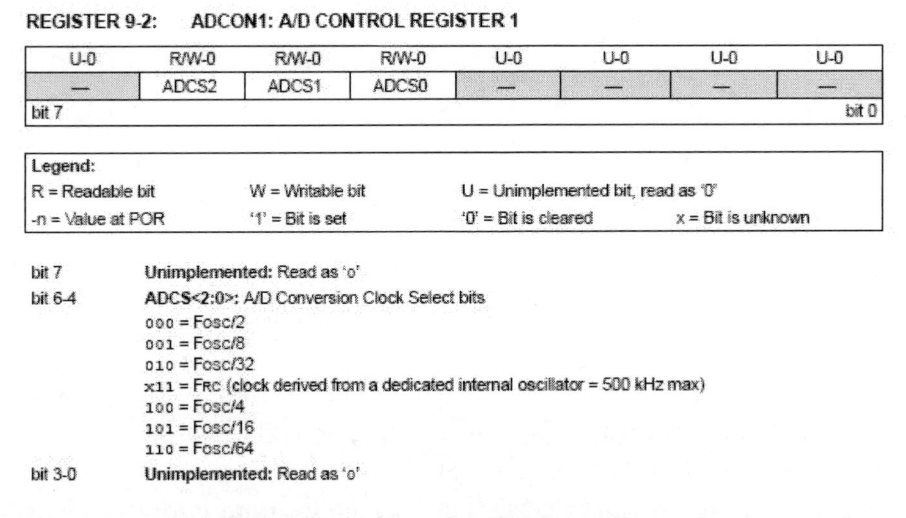

REGISTER 9-2: ADCON1: A/D CONTROL REGISTER 1

U-0	R/W-0	R/W-0	R/W-0	U-0	U-0	U-0	U-0
—	ADCS2	ADCS1	ADCS0	—	—	—	—
bit 7							bit 0

Legend:			
R = Readable bit	W = Writable bit	U = Unimplemented bit, read as '0'	
-n = Value at POR	'1' = Bit is set	'0' = Bit is cleared	x = Bit is unknown

bit 7 **Unimplemented:** Read as '0'

bit 6-4 **ADCS<2:0>:** A/D Conversion Clock Select bits
000 = Fosc/2
001 = Fosc/8
010 = Fosc/32
x11 = FRC (clock derived from a dedicated internal oscillator = 500 kHz max)
100 = Fosc/4
101 = Fosc/16
110 = Fosc/64

bit 3-0 **Unimplemented:** Read as '0'

Figure 15-6: ADCON1 Register

Ok, the A/D register is setup so we finish off the initial part of main loop by clearing the PORTC register and then setting all of PORTC to outputs. This makes sure the LEDs start in the off state.

```
PORTC = 0;                //Clear PortC
TRISC = 0;                //All PortC is output
```

The action part of the main loop starts with a familiar *while* statement.

```
while(1==1)              //Loop Forever
{
```

The A/D conversion is started by setting Bit1 of the ADCON0 register. The PICC-Lite compiler has already defined a nickname for that bit called GODONE. This is part of the PIC16F690 header file that the pic.h calls in. The command line below sets the GODONE bit to one which starts the A/D conversion. The A/D port will start converting the voltage of the AN0 pin which is tied to the potentiometer on the development board.

```
GODONE = 1; //Start A/D process
```

Now the program waits for the GODONE bit to indicate the A/D conversion is complete by changing to a zero. We test for that with a *while* statement that does nothing while the bit is set to a one.

```
        while (GODONE ==1) //wait for A/D to finish
        {
        }
```

When the conversion is complete the GODONE bit changes to zero and the *while* statement releases program control to the next line. When the A/D conversion is complete the converted value is left shifted and the upper eight bits are stored in the ADRESH register. We want to save that in a variable so we don't lose that data. Actually it will stay there until a new A/D conversion is performed but in future programs you may want to do more A/D conversions so its good to place the result in a specific variable. The line below does that.

```
advalue = ADRESH;        //Store A/D result in variable ADVALUE
```

The next set of command lines should be understandable to you by now as we are just testing the value of the variable "advalue" against some fixed values. Each *if* statement gets tested which is different then *switch-case* which has a *break* statement to jump out. When the *if* statement is true, the LEDs are set to those binary constants established at the beginning of the program. Each *if* statement compares the "advalue" variable against a constant value and more than one *if* statement can be true. The last one that is true is the one that wins out and controls the LEDs.

```
if (advalue <60)  //Test if A/D value is less than 60 decimal
{
PORTC = DS4_on;   //Less than 60, light DS4 LED
}
if (advalue <120) //Test if A/D value less than 120 decimal
{
PORTC = DS3_on;   //Less than 120, light DS3 LED
}
if (advalue <180) //Test if A/D value less than 180
{
PORTC = DS2_on;   //Less than 180, light DS2 LED
}
if (advalue < 240) //Test if A/D value less than 240
{
PORTC = DS1_on;   //Less than 240, light DS1 LED
}
```

If the value of "advalue" is larger than 240 which means the potentiometer is cranked all the way to the end of its travel, all the LEDs will be set off by clearing the PORTC register.

```
if (advalue >= 240)//Test if A/D value equal to
                   // or greater than 240
{
PORTC = 0;         //Set all LEDs off
}
```

The final lines of the program just close out the curly brackets.

```
}      //End While
}      //End Main
```

Next Steps

The most obvious next step is to change the values of the *if* statements to make the LEDs come on at a different position of the potentiometer but that will not really be noticeable. What I recommend is to change the LED binary settings at the top of the program to light more than one LED at a time to make it look more like a bar graph display rather than a single moving LED.

Chapter 16 – Conclusion

This concludes the examples and topics this book will cover and I hope you feel a little more comfortable by now writing software with a C compiler. There is a lot more you can do beyond the simple projects presented. I hope to add more books with projects that take you to the next step but overall learning to program is just like any other talent, it takes practice. The whole point of this book was to take you step by step through the basics of programming in C so you can go further and hopefully understand the programs written by other people. You should be able to borrow sections of code from other programmers that you can modify to fit your needs. You can also try to find libraries of functions that are pre-written for you. I know the PICC compiler, which is the full functional version of the PICC-Lite compiler has a whole list of library functions. There are other C compilers you can check out as well including the C18 compiler from Microchip that is written to work with the PIC18 family of parts.

The PICkit 2 hardware used in this book is dedicated to the smaller 8, 14 and 20 pin parts but there are several different development boards available that will connect to the same PICkit 2 programmer. Therefore all the setup you did to get the hardware working will also work when using other parts and compilers with the Microchip PIC Micocontrollers.

This is by no means a complete end all book on the C programming language but it should have helped you understand how to program an embedded microcontroller with the C language. Please give me your feedback or send any questions to my email at chuck@elproducts.com. You can visit my website for any corrections to this book as well at www.elproducts.com. All the files can be downloaded from there as well.

Finally, thanks for purchasing my book. It's your contributions that make this book financially worth my time to keep writing and helping others enjoy programming embedded controllers as much as I do. I plan to follow this book with more books with more advanced projects.

Appendix A - PIC16F690 Configuration Options

REGISTER 14-1: CONFIG: CONFIGURATION WORD REGISTER

Reserved	Reserved	FCMEN	IESO	BOREN1[1]	BOREN0[1]	$\overline{CPD}$[2]
bit 13						bit 7

$\overline{CP}$[3]	MCLRE[4]	$\overline{PWRTE}$	WDTE	FOSC2	FOSC1	FOSC0
bit 6						bit 0

Legend:			
R = Readable bit	W = Writable bit	P = Programmable'	U = Unimplemented bit, read as '0'
-n = Value at POR	'1' = Bit is set	'0' = Bit is cleared	x = Bit is unknown

bit 13-12 **Reserved:** Reserved bits. Do Not Use.

bit 11 **FCMEN:** Fail-Safe Clock Monitor Enabled bit
1 = Fail-Safe Clock Monitor is enabled
0 = Fail-Safe Clock Monitor is disabled

bit 10 **IESO:** Internal External Switchover bit
1 = Internal External Switchover mode is enabled
0 = Internal External Switchover mode is disabled

bit 9-8 **BOREN<1:0>:** Brown-out Reset Selection bits[1]
11 = BOR enabled
10 = BOR enabled during operation and disabled in Sleep
01 = BOR controlled by SBOREN bit of the PCON register
00 = BOR disabled

bit 7 **$\overline{CPD}$:** Data Code Protection bit[2]
1 = Data memory code protection is disabled
0 = Data memory code protection is enabled

bit 6 **$\overline{CP}$:** Code Protection bit[2]
1 = Program memory code protection is disabled
0 = Program memory code protection is enabled

bit 5 **MCLRE:** $\overline{MCLR}$ Pin Function Select bit[3]
1 = $\overline{MCLR}$ pin function is $\overline{MCLR}$
0 = $\overline{MCLR}$ pin function is digital input, $\overline{MCLR}$ internally tied to VDD

bit 4 **$\overline{PWRTE}$:** Power-up Timer Enable bit
1 = PWRT disabled
0 = PWRT enabled

bit 3 **WDTE:** Watchdog Timer Enable bit
1 = WDT enabled
0 = WDT disabled

bit 2-0 **FOSC<2:0>:** Oscillator Selection bits
111 = RC oscillator: CLKOUT function on RA4/OSC2/CLKOUT pin, RC on RA5/OSC1/CLKIN
110 = RCIO oscillator: I/O function on RA4/OSC2/CLKOUT pin, RC on RA5/OSC1/CLKIN
101 = INTOSC oscillator: CLKOUT function on RA4/OSC2/CLKOUT pin, I/O function on RA5/OSC1/CLKIN
100 = INTOSCIO oscillator: I/O function on RA4/OSC2/CLKOUT pin, I/O function on RA5/OSC1/CLKIN
011 = EC: I/O function on RA4/OSC2/CLKOUT pin, CLKIN on RA5/OSC1/CLKIN
010 = HS oscillator: High-speed crystal/resonator on RA4/OSC2/CLKOUT and RA5/OSC1/CLKIN
001 = XT oscillator: Crystal/resonator on RA4/OSC2/CLKOUT and RA5/OSC1/CLKIN
000 = LP oscillator: Low-power crystal on RA4/OSC2/CLKOUT and RA5/OSC1/CLKIN

Note 1: Enabling Brown-out Reset does not automatically enable Power-up Timer.
2: The entire data EEPROM will be erased when the code protection is turned off.
3: The entire program memory will be erased when the code protection is turned off.
4: When $\overline{MCLR}$ is asserted in INTOSC or RC mode, the internal clock oscillator is disabled.

PICC-Lite Configuration Definitions (found in pic16F685.h file)

```
// Configuration Mask Definitions
#define CONFIG_ADDR      0x2007

// Oscillator
#define EXTCLK       0x3FFF    // External RC Clockout
#define EXTIO        0x3FFE    // External RC No Clock
#define INTCLK       0x3FFD    // Internal RC Clockout
#define INTIO        0x3FFC    // Internal RC No Clock
#define EC      0x3FFB    // EC
#define HS      0x3FFA    // HS
#define XT      0x3FF9    // XT
#define LP      0x3FF8    // LP

// Watchdog Timer
#define WDTEN        0x3FFF    // On
#define WDTDIS       0x3FF7    // Off

// Power Up Timer
#define PWRTDIS      0x3FFF    // Off
#define PWRTEN       0x3FEF    // On

// Master Clear Enable
#define MCLREN       0x3FFF    // MCLR function is enabled
#define MCLRDIS      0x3FDF    // MCLR functions as IO

// Code Protect
#define UNPROTECT    0x3FFF    // Code is not protected
#define CP      0x3FBF    // Code is protected

// Data EE Read Protect
#define UNPROTECT    0x3FFF    // Do not read protect EEPROM data
#define CPD          0x3F7F    // Read protect EEPROM data

// Brown Out Detect
#define BORDIS       0x3CFF    // BOD and SBOREN disabled
#define SWBOREN           0x3DFF    // SBOREN controls BOR
                                    //function (Software control)
```

```
#define BORXSLP         0x3EFF   // BOD enabled in run, disabled
                                 //in sleep, SBOREN disabled
#define BOREN       0x3FFF   // BOD Enabled, SBOREN Disabled

// Internal External Switch Over Mode
#define IESOEN      0x3FFF   // Enabled
#define IESODIS     0x3BFF   // Disabled

// Monitor Clock Fail-safe
#define FCMEN       0x3FFF   // Enabled
#define FCMDIS      0x37FF   // Disabled
```

Appendix B – pic.h and pic16f685.h include files

Section of pic.h file include for PIC16F690

```
#if defined(_16F631)|| defined(_16F677)|| defined(_16F685)||\
  defined(_16F687)|| defined(_16F689)|| defined(_16F690)
      #include      <pic16f685.h>
#endif
```

---

pic16f685.h file called from the pic.h file above.

```
#ifndef _HTC_H_
#warning Header file pic16f685.h included directly. Use #include <htc.h> instead.
#endif


/* header file for the MICROCHIP PIC microcontrollers
        PIC16F631
        PIC16F677
        PIC16F685
        PIC16F687
        PIC16F689
        PIC16F690
*/

#ifndef __PIC16F685_H
#define __PIC16F685_H

// Special function register definitions

static volatile    unsigned char    TMR0      @ 0x001;
static volatile    unsigned char    PCL       @ 0x002;
static volatile    unsigned char    STATUS          @ 0x003;
static          unsigned char FSR       @ 0x004;
static volatile    unsigned char    PORTA           @ 0x005;
static volatile    unsigned char    PORTB           @ 0x006;
static volatile    unsigned char    PORTC           @ 0x007;
static volatile    unsigned char    PCLATH          @ 0x00A;
static volatile    unsigned char    INTCON          @ 0x00B;
static volatile    unsigned char    PIR1      @ 0x00C;
```

```
static volatile     unsigned char     PIR2          @ 0x00D;
static volatile     unsigned char     TMR1L              @ 0x00E;
static volatile     unsigned char     TMR1H              @ 0x00F;
static          unsigned char T1CON             @ 0x010;
#if defined(_16F685) || defined(_16F690)
static volatile     unsigned char     TMR2          @ 0x011;
static          unsigned char T2CON             @ 0x012;
#endif
#if defined(_16F677) || defined(_16F687) || defined(_16F689) || defined(_16F690)
static volatile     unsigned char     SSPBUF             @ 0x013;
static volatile     unsigned char     SSPCON             @ 0x014;
#endif
#if defined(_16F685) || defined(_16F690)
static volatile     unsigned char     CCPR1L             @ 0x015;
static volatile     unsigned char     CCPR1H             @ 0x016;
static volatile     unsigned char     CCP1CON            @ 0x017;
#endif
#if defined(_16F687) || defined(_16F689) || defined(_16F690)
static volatile     unsigned char     RCSTA              @ 0x018;
static volatile     unsigned char     TXREG              @ 0x019;
static volatile     unsigned char     RCREG              @ 0x01A;
#endif
#if defined(_16F685) || defined(_16F690)
static volatile     unsigned char     PWM1CON            @ 0x01C;
static volatile     unsigned char     ECCPAS             @ 0x01D;
#endif
#if defined(_16F677) || defined(_16F685) || defined(_16F687) || defined(_16F689) ||
defined(_16F690)
static volatile     unsigned char     ADRESH             @ 0x01E;
static volatile     unsigned char     ADCON0             @ 0x01F;
#endif
static       bank1 unsigned char      OPTION             @ 0x081;
static volatile bank1 unsigned char   TRISA         @ 0x085;
static volatile bank1 unsigned char   TRISB         @ 0x086;
static volatile bank1 unsigned char   TRISC         @ 0x087;
static       bank1 unsigned char      PIE1          @ 0x08C;
static       bank1 unsigned char      PIE2          @ 0x08D;
static volatile bank1 unsigned char   PCON          @ 0x08E;
static volatile bank1 unsigned char   OSCCON             @ 0x08F;
static       bank1 unsigned char      OSCTUNE            @ 0x090;
#if defined(_16F685) || defined(_16F690)
static       bank1 unsigned char      PR2           @ 0x092;
#endif
#if defined(_16F677) || defined(_16F687) || defined(_16F689) || defined(_16F690)
```

```
static         bank1 unsigned char       SSPADD              @ 0x093;
// Alternate definition
static         bank1 unsigned char       SSPMSK              @ 0x093;
static volatile bank1 unsigned char      SSPSTAT             @ 0x094;
#endif
static         bank1 unsigned char       WPUA       @ 0x095;
static         bank1 unsigned char       IOCA       @ 0x096;
static volatile bank1 unsigned char      WDTCON              @ 0x097;
#if defined(_16F687) || defined(_16F689) || defined(_16F690)
static volatile bank1 unsigned char      TXSTA               @ 0x098;
static         bank1 unsigned char       SPBRG               @ 0x099;
static         bank1 unsigned char       SPBRGH              @ 0x09A;
static volatile bank1 unsigned char      BAUDCTL             @ 0x09B;
#endif
#if defined(_16F677) || defined(_16F685) || defined(_16F687) || defined(_16F689) ||
defined(_16F690)
static volatile bank1 unsigned char      ADRESL              @ 0x09E;
static         bank1 unsigned char       ADCON1              @ 0x09F;
#endif
static volatile bank2 unsigned char      EEDAT               @ 0x10C;
/* Alternate definition */
static volatile bank2 unsigned char      EEDATA              @ 0x10C;
/* Alternate definition */
static volatile bank2 unsigned char      EEDATL              @ 0x10C;
static         bank2 unsigned char       EEADR               @ 0x10D;
/* Alternate definition */
static         bank2 unsigned char       EEADRL              @ 0x10D;
#if defined(_16F685) || defined(_16F689) || defined(_16F690)
static volatile bank2 unsigned char      EEDATH              @ 0x10E;
static         bank2 unsigned char       EEADRH              @ 0x10F;
#endif
static         bank2 unsigned char       WPUB       @ 0x115;
static         bank2 unsigned char       IOCB       @ 0x116;
static         bank2 unsigned char       VRCON               @ 0x118;
static volatile bank2 unsigned char      CM1CON0             @ 0x119;
static volatile bank2 unsigned char      CM2CON0             @ 0x11A;
static volatile bank2 unsigned char      CM2CON1             @ 0x11B;
static         bank2 unsigned char       ANSEL               @ 0x11E;
#if defined(_16F677) || defined(_16F685) || defined(_16F687) || defined(_16F689) ||
defined(_16F690)
static         bank2 unsigned char       ANSELH              @ 0x11F;
#endif
static volatile bank3 unsigned char      EECON1              @ 0x18C;
static volatile bank3 unsigned char      EECON2              @ 0x18D;
```

```
#if defined(_16F685) || defined(_16F690)
static          bank3 unsigned char      PSTRCON          @ 0x19D;
#endif
static volatile bank3 unsigned char      SRCON            @ 0x19E;


/* Definitions for STATUS register */
static volatile     bit   CARRY              @ ((unsigned)&STATUS*8)+0;
static volatile     bit   DC          @ ((unsigned)&STATUS*8)+1;
static volatile     bit   ZERO        @ ((unsigned)&STATUS*8)+2;
static volatile     bit   PD          @ ((unsigned)&STATUS*8)+3;
static volatile     bit   TO          @ ((unsigned)&STATUS*8)+4;
static          bit       RP0         @ ((unsigned)&STATUS*8)+5;
static          bit       RP1         @ ((unsigned)&STATUS*8)+6;
static          bit       IRP         @ ((unsigned)&STATUS*8)+7;

/* Definitions for PORTA register */
static volatile     bit   RA0         @ ((unsigned)&PORTA*8)+0;
static volatile     bit   RA1         @ ((unsigned)&PORTA*8)+1;
static volatile     bit   RA2         @ ((unsigned)&PORTA*8)+2;
static volatile     bit   RA3         @ ((unsigned)&PORTA*8)+3;
static volatile     bit   RA4         @ ((unsigned)&PORTA*8)+4;
static volatile     bit   RA5         @ ((unsigned)&PORTA*8)+5;

/* Definitions for PORTB register */
static volatile     bit   RB4         @ ((unsigned)&PORTB*8)+4;
static volatile     bit   RB5         @ ((unsigned)&PORTB*8)+5;
static volatile     bit   RB6         @ ((unsigned)&PORTB*8)+6;
static volatile     bit   RB7         @ ((unsigned)&PORTB*8)+7;

/* Definitions for PORTC register */
static volatile     bit   RC0         @ ((unsigned)&PORTC*8)+0;
static volatile     bit   RC1         @ ((unsigned)&PORTC*8)+1;
static volatile     bit   RC2         @ ((unsigned)&PORTC*8)+2;
static volatile     bit   RC3         @ ((unsigned)&PORTC*8)+3;
static volatile     bit   RC4         @ ((unsigned)&PORTC*8)+4;
static volatile     bit   RC5         @ ((unsigned)&PORTC*8)+5;
static volatile     bit   RC6         @ ((unsigned)&PORTC*8)+6;
static volatile     bit   RC7         @ ((unsigned)&PORTC*8)+7;

/* Definitions for INTCON register */
static volatile     bit   RABIF       @ ((unsigned)&INTCON*8)+0;
// Alternate definition for backward compatibility
static volatile     bit   RBIF        @ ((unsigned)&INTCON*8)+0;
```

```
static volatile     bit  INTF        @ ((unsigned)&INTCON*8)+1;
static volatile     bit  T0IF        @ ((unsigned)&INTCON*8)+2;
static          bit      RABIE             @ ((unsigned)&INTCON*8)+3;
// Alternate definition for backward compatibility
static          bit      RBIE        @ ((unsigned)&INTCON*8)+3;
static          bit      INTE        @ ((unsigned)&INTCON*8)+4;
static          bit      T0IE        @ ((unsigned)&INTCON*8)+5;
static          bit      PEIE        @ ((unsigned)&INTCON*8)+6;
static          bit      GIE         @ ((unsigned)&INTCON*8)+7;

/* Definitions for PIR1 register */
static volatile     bit  TMR1IF            @ ((unsigned)&PIR1*8)+0;
#if defined(_16F685) || defined(_16F690)
static volatile     bit  TMR2IF            @ ((unsigned)&PIR1*8)+1;
static volatile     bit  CCP1IF            @ ((unsigned)&PIR1*8)+2;
#endif
#if defined(_16F687) || defined(_16F689) || defined(_16F690)
static volatile     bit  SSPIF       @ ((unsigned)&PIR1*8)+3;
static volatile     bit  TXIF        @ ((unsigned)&PIR1*8)+4;
static volatile     bit  RCIF        @ ((unsigned)&PIR1*8)+5;
#endif
#if defined(_16F677) || defined(_16F685) || defined(_16F687) || defined(_16F689) ||
defined(_16F690)
static volatile     bit  ADIF        @ ((unsigned)&PIR1*8)+6;
#endif

/* Definitions for PIR2 register */
static volatile     bit  EEIF        @ ((unsigned)&PIR2*8)+4;
static volatile     bit  C1IF        @ ((unsigned)&PIR2*8)+5;
static volatile     bit  C2IF        @ ((unsigned)&PIR2*8)+6;
static volatile     bit  OSFIF       @ ((unsigned)&PIR2*8)+7;

/* Definitions for T1CON register */
static          bit      TMR1ON            @ ((unsigned)&T1CON*8)+0;
static          bit      TMR1CS            @ ((unsigned)&T1CON*8)+1;
static          bit      T1SYNC            @ ((unsigned)&T1CON*8)+2;
static          bit      T1OSCEN           @ ((unsigned)&T1CON*8)+3;
static          bit      T1CKPS0           @ ((unsigned)&T1CON*8)+4;
static          bit      T1CKPS1           @ ((unsigned)&T1CON*8)+5;
static          bit      TMR1GE            @ ((unsigned)&T1CON*8)+6;
static          bit      T1GINV            @ ((unsigned)&T1CON*8)+7;

#if defined(_16F685) || defined(_16F690)
/* Definitions for T2CON register */
```

```
static          bit     T2CKPS0           @ ((unsigned)&T2CON*8)+0;
static          bit     T2CKPS1           @ ((unsigned)&T2CON*8)+1;
static          bit     TMR2ON            @ ((unsigned)&T2CON*8)+2;
static          bit     TOUTPS0           @ ((unsigned)&T2CON*8)+3;
static          bit     TOUTPS1           @ ((unsigned)&T2CON*8)+4;
static          bit     TOUTPS2           @ ((unsigned)&T2CON*8)+5;
static          bit     TOUTPS3           @ ((unsigned)&T2CON*8)+6;
#endif

#if defined(_16F677) || defined(_16F687) || defined(_16F689) || defined(_16F690)
/* Definitions for SSPCON register */
static          bit     SSPM0             @ ((unsigned)&SSPCON*8)+0;
static          bit     SSPM1             @ ((unsigned)&SSPCON*8)+1;
static          bit     SSPM2             @ ((unsigned)&SSPCON*8)+2;
static          bit     SSPM3             @ ((unsigned)&SSPCON*8)+3;
static          bit     CKP       @ ((unsigned)&SSPCON*8)+4;
static          bit     SSPEN             @ ((unsigned)&SSPCON*8)+5;
static volatile     bit SSPOV             @ ((unsigned)&SSPCON*8)+6;
static volatile     bit WCOL      @ ((unsigned)&SSPCON*8)+7;
#endif

#if defined(_16F685) || defined(_16F690)
/* Definitions for CCP1CON register */
static          bit     CCP1M0            @ ((unsigned)&CCP1CON*8)+0;
static          bit     CCP1M1            @ ((unsigned)&CCP1CON*8)+1;
static          bit     CCP1M2            @ ((unsigned)&CCP1CON*8)+2;
static          bit     CCP1M3            @ ((unsigned)&CCP1CON*8)+3;
static          bit     DC1B0             @ ((unsigned)&CCP1CON*8)+4;
static          bit     DC1B1             @ ((unsigned)&CCP1CON*8)+5;
static          bit     P1M0      @ ((unsigned)&CCP1CON*8)+6;
static          bit     P1M1      @ ((unsigned)&CCP1CON*8)+7;
#endif

#if defined(_16F687) || defined(_16F689) || defined(_16F690)
/* Definitions for RCSTA register */
static volatile     bit RX9D      @ ((unsigned)&RCSTA*8)+0;
static volatile     bit OERR      @ ((unsigned)&RCSTA*8)+1;
static volatile     bit FERR      @ ((unsigned)&RCSTA*8)+2;
static          bit     ADDEN             @ ((unsigned)&RCSTA*8)+3;
static          bit     CREN      @ ((unsigned)&RCSTA*8)+4;
static          bit     SREN      @ ((unsigned)&RCSTA*8)+5;
static          bit     RX9       @ ((unsigned)&RCSTA*8)+6;
static          bit     SPEN      @ ((unsigned)&RCSTA*8)+7;
#endif
```

```
#if defined(_16F685) || defined(_16F690)
/* Definitions for PWM1CON register */
static volatile     bit  PDC0          @ ((unsigned)&PWM1CON*8)+0;
static volatile     bit  PDC1          @ ((unsigned)&PWM1CON*8)+1;
static volatile     bit  PDC2          @ ((unsigned)&PWM1CON*8)+2;
static volatile     bit  PDC3          @ ((unsigned)&PWM1CON*8)+3;
static volatile     bit  PDC4          @ ((unsigned)&PWM1CON*8)+4;
static volatile     bit  PDC5          @ ((unsigned)&PWM1CON*8)+5;
static volatile     bit  PDC6          @ ((unsigned)&PWM1CON*8)+6;
static volatile     bit  PRSEN                @ ((unsigned)&PWM1CON*8)+7;

/* Definitions for ECCPAS register */
static          bit     PSSBD0              @ ((unsigned)&ECCPAS*8)+0;
static          bit     PSSBD1              @ ((unsigned)&ECCPAS*8)+1;
static          bit     PSSAC0              @ ((unsigned)&ECCPAS*8)+2;
static          bit     PSSAC1              @ ((unsigned)&ECCPAS*8)+3;
static          bit     ECCPAS0             @ ((unsigned)&ECCPAS*8)+4;
static          bit     ECCPAS1             @ ((unsigned)&ECCPAS*8)+5;
static          bit     ECCPAS2             @ ((unsigned)&ECCPAS*8)+6;
static volatile   bit   ECCPASE             @ ((unsigned)&ECCPAS*8)+7;
#endif

#if defined(_16F677) || defined(_16F685) || defined(_16F687) || defined(_16F689) ||
defined(_16F690)
/* Definitions for ADCON0 register */
static          bit     ADON          @ ((unsigned)&ADCON0*8)+0;
static volatile   bit   GODONE              @ ((unsigned)&ADCON0*8)+1;
static          bit     CHS0          @ ((unsigned)&ADCON0*8)+2;
static          bit     CHS1          @ ((unsigned)&ADCON0*8)+3;
static          bit     CHS2          @ ((unsigned)&ADCON0*8)+4;
static          bit     CHS3          @ ((unsigned)&ADCON0*8)+5;
static          bit     VCFG          @ ((unsigned)&ADCON0*8)+6;
static          bit     ADFM          @ ((unsigned)&ADCON0*8)+7;
#endif

/* Definitions for OPTION register */
static      bank1 bit PS0           @ ((unsigned)&OPTION*8)+0;
static      bank1 bit PS1           @ ((unsigned)&OPTION*8)+1;
static      bank1 bit PS2           @ ((unsigned)&OPTION*8)+2;
static      bank1 bit PSA           @ ((unsigned)&OPTION*8)+3;
static      bank1 bit T0SE          @ ((unsigned)&OPTION*8)+4;
static      bank1 bit T0CS          @ ((unsigned)&OPTION*8)+5;
static      bank1 bit INTEDG              @ ((unsigned)&OPTION*8)+6;
```

```
static      bank1 bit  RABPU              @ ((unsigned)&OPTION*8)+7;
// Alternate definition for backward compatibility
static      bank1 bit  RBPU         @ ((unsigned)&OPTION*8)+7;

/* Definitions for TRISA register */
static      bank1 bit  TRISA0             @ ((unsigned)&TRISA*8)+0;
static      bank1 bit  TRISA1             @ ((unsigned)&TRISA*8)+1;
static      bank1 bit  TRISA2             @ ((unsigned)&TRISA*8)+2;
static      bank1 bit  TRISA3             @ ((unsigned)&TRISA*8)+3;
static      bank1 bit  TRISA4             @ ((unsigned)&TRISA*8)+4;
static      bank1 bit  TRISA5             @ ((unsigned)&TRISA*8)+5;

/* Definitions for TRISB register */
static volatile bank1 bit      TRISB4            @ ((unsigned)&TRISB*8)+4;
static volatile bank1 bit      TRISB5            @ ((unsigned)&TRISB*8)+5;
static volatile bank1 bit      TRISB6            @ ((unsigned)&TRISB*8)+6;
static volatile bank1 bit      TRISB7            @ ((unsigned)&TRISB*8)+7;

/* Definitions for TRISC register */
static volatile bank1 bit      TRISC0            @ ((unsigned)&TRISC*8)+0;
static volatile bank1 bit      TRISC1            @ ((unsigned)&TRISC*8)+1;
static volatile bank1 bit      TRISC2            @ ((unsigned)&TRISC*8)+2;
static volatile bank1 bit      TRISC3            @ ((unsigned)&TRISC*8)+3;
static volatile bank1 bit      TRISC4            @ ((unsigned)&TRISC*8)+4;
static volatile bank1 bit      TRISC5            @ ((unsigned)&TRISC*8)+5;
static volatile bank1 bit      TRISC6            @ ((unsigned)&TRISC*8)+6;
static volatile bank1 bit      TRISC7            @ ((unsigned)&TRISC*8)+7;

/* Definitions for PIE1 register */
static      bank1 bit  TMR1IE             @ ((unsigned)&PIE1*8)+0;
#if defined(_16F685) || defined(_16F690)
static      bank1 bit  TMR2IE             @ ((unsigned)&PIE1*8)+1;
static      bank1 bit  CCP1IE             @ ((unsigned)&PIE1*8)+2;
#endif
#if defined(_16F687) || defined(_16F689) || defined(_16F690)
static      bank1 bit  SSPIE        @ ((unsigned)&PIE1*8)+3;
static      bank1 bit  TXIE         @ ((unsigned)&PIE1*8)+4;
static      bank1 bit  RCIE         @ ((unsigned)&PIE1*8)+5;
#endif
#if defined(_16F677) || defined(_16F685) || defined(_16F687) || defined(_16F689) ||
defined(_16F690)
static      bank1 bit  ADIE         @ ((unsigned)&PIE1*8)+6;
#endif
```

```
/* Definitions for PIE2 register */
static        bank1 bit  EEIE          @ ((unsigned)&PIE2*8)+4;
static        bank1 bit  C1IE          @ ((unsigned)&PIE2*8)+5;
static        bank1 bit  C2IE          @ ((unsigned)&PIE2*8)+6;
static        bank1 bit  OSFIE         @ ((unsigned)&PIE2*8)+7;

/* Definitions for PCON register */
static volatile bank1 bit    BOR            @ ((unsigned)&PCON*8)+0;
static volatile bank1 bit    POR            @ ((unsigned)&PCON*8)+1;
static        bank1 bit  SBOREN             @ ((unsigned)&PCON*8)+4;
static        bank1 bit  ULPWUE             @ ((unsigned)&PCON*8)+5;

/* Definitions for OSCCON register */
static        bank1 bit  SCS           @ ((unsigned)&OSCCON*8)+0;
static volatile bank1 bit    LTS            @ ((unsigned)&OSCCON*8)+1;
static volatile bank1 bit    HTS            @ ((unsigned)&OSCCON*8)+2;
static volatile bank1 bit    OSTS           @ ((unsigned)&OSCCON*8)+3;
static        bank1 bit  IRCF0         @ ((unsigned)&OSCCON*8)+4;
static        bank1 bit  IRCF1         @ ((unsigned)&OSCCON*8)+5;
static        bank1 bit  IRCF2         @ ((unsigned)&OSCCON*8)+6;

/* Definitions for OSCTUNE register */
static        bank1 bit  TUN0          @ ((unsigned)&OSCTUNE*8)+0;
static        bank1 bit  TUN1          @ ((unsigned)&OSCTUNE*8)+1;
static        bank1 bit  TUN2          @ ((unsigned)&OSCTUNE*8)+2;
static        bank1 bit  TUN3          @ ((unsigned)&OSCTUNE*8)+3;
static        bank1 bit  TUN4          @ ((unsigned)&OSCTUNE*8)+4;

#if defined(_16F677) || defined(_16F687) || defined(_16F689) || defined(_16F690)
/* Definitions for SSPSTAT register */
static volatile bank1 bit    BF             @ ((unsigned)&SSPSTAT*8)+0;
static volatile bank1 bit    UA             @ ((unsigned)&SSPSTAT*8)+1;
static volatile bank1 bit    RW             @ ((unsigned)&SSPSTAT*8)+2;
static volatile bank1 bit    START               @ ((unsigned)&SSPSTAT*8)+3;
static volatile bank1 bit    STOP           @ ((unsigned)&SSPSTAT*8)+4;
static volatile bank1 bit    DA             @ ((unsigned)&SSPSTAT*8)+5;
static        bank1 bit  CKE           @ ((unsigned)&SSPSTAT*8)+6;
static        bank1 bit  SMP           @ ((unsigned)&SSPSTAT*8)+7;
#endif

/* Definitions for WPUA register */
static        bank1 bit  WPUA0              @ ((unsigned)&WPUA*8)+0;
static        bank1 bit  WPUA1              @ ((unsigned)&WPUA*8)+1;
static        bank1 bit  WPUA2              @ ((unsigned)&WPUA*8)+2;
```

```
static      bank1 bit  WPUA4        @ ((unsigned)&WPUA*8)+4;
static      bank1 bit  WPUA5        @ ((unsigned)&WPUA*8)+5;

/* Definitions for IOCA register */
static      bank1 bit  IOCA0    @ ((unsigned)&IOCA*8)+0;
static      bank1 bit  IOCA1    @ ((unsigned)&IOCA*8)+1;
static      bank1 bit  IOCA2    @ ((unsigned)&IOCA*8)+2;
static      bank1 bit  IOCA3    @ ((unsigned)&IOCA*8)+3;
static      bank1 bit  IOCA4    @ ((unsigned)&IOCA*8)+4;
static      bank1 bit  IOCA5    @ ((unsigned)&IOCA*8)+5;

/* Definitions for WDTCON register */
static      bank1 bit  SWDTEN       @ ((unsigned)&WDTCON*8)+0;
static      bank1 bit  WDTPS0       @ ((unsigned)&WDTCON*8)+1;
static      bank1 bit  WDTPS1       @ ((unsigned)&WDTCON*8)+2;
static      bank1 bit  WDTPS2       @ ((unsigned)&WDTCON*8)+3;
static      bank1 bit  WDTPS3       @ ((unsigned)&WDTCON*8)+4;

#if defined(_16F687) || defined(_16F689) || defined(_16F690)
/* Definitions for TXSTA register */
static volatile bank1 bit   TX9D    @ ((unsigned)&TXSTA*8)+0;
static volatile bank1 bit   TRMT    @ ((unsigned)&TXSTA*8)+1;
static      bank1 bit  BRGH     @ ((unsigned)&TXSTA*8)+2;
static      bank1 bit  SENDB        @ ((unsigned)&TXSTA*8)+3;
static      bank1 bit  SYNC     @ ((unsigned)&TXSTA*8)+4;
static      bank1 bit  TXEN     @ ((unsigned)&TXSTA*8)+5;
static      bank1 bit  TX9      @ ((unsigned)&TXSTA*8)+6;
static      bank1 bit  CSRC     @ ((unsigned)&TXSTA*8)+7;

/* Definitions for SPBRG register */
static      bank1 bit  BRG0     @ ((unsigned)&SPBRG*8)+0;
static      bank1 bit  BRG1     @ ((unsigned)&SPBRG*8)+1;
static      bank1 bit  BRG2     @ ((unsigned)&SPBRG*8)+2;
static      bank1 bit  BRG3     @ ((unsigned)&SPBRG*8)+3;
static      bank1 bit  BRG4     @ ((unsigned)&SPBRG*8)+4;
static      bank1 bit  BRG5     @ ((unsigned)&SPBRG*8)+5;
static      bank1 bit  BRG6     @ ((unsigned)&SPBRG*8)+6;
static      bank1 bit  BRG7     @ ((unsigned)&SPBRG*8)+7;

/* Definitions for SPBRGH register */
static      bank1 bit  BRG8     @ ((unsigned)&SPBRGH*8)+0;
static      bank1 bit  BRG9     @ ((unsigned)&SPBRGH*8)+1;
static      bank1 bit  BRG10        @ ((unsigned)&SPBRGH*8)+2;
static      bank1 bit  BRG11        @ ((unsigned)&SPBRGH*8)+3;
```

```
static        bank1 bit  BRG12                @ ((unsigned)&SPBRGH*8)+4;
static        bank1 bit  BRG13                @ ((unsigned)&SPBRGH*8)+5;
static        bank1 bit  BRG14                @ ((unsigned)&SPBRGH*8)+6;
static        bank1 bit  BRG15                @ ((unsigned)&SPBRGH*8)+7;

/* Definitions for BAUDCTL register */
static volatile bank1 bit       ABDEN              @ ((unsigned)&BAUDCTL*8)+0;
static volatile bank1 bit       WUE           @ ((unsigned)&BAUDCTL*8)+1;
static        bank1 bit  BRG16                @ ((unsigned)&BAUDCTL*8)+3;
static volatile bank1 bit       SCKP          @ ((unsigned)&BAUDCTL*8)+4;
static volatile bank1 bit       RCIDL              @ ((unsigned)&BAUDCTL*8)+6;
static volatile bank1 bit       ABDOVF             @ ((unsigned)&BAUDCTL*8)+7;
#endif

#if defined(_16F677) || defined(_16F685) || defined(_16F687) || defined(_16F689) ||
defined(_16F690)
/* Definitions for ADCON1 register */
static        bank1 bit  ADCS0                @ ((unsigned)&ADCON1*8)+4;
static        bank1 bit  ADCS1                @ ((unsigned)&ADCON1*8)+5;
static        bank1 bit  ADCS2                @ ((unsigned)&ADCON1*8)+6;
#endif

/* Definitions for WPUB register */
static        bank2 bit  WPUB4                @ ((unsigned)&WPUB*8)+4;
static        bank2 bit  WPUB5                @ ((unsigned)&WPUB*8)+5;
static        bank2 bit  WPUB6                @ ((unsigned)&WPUB*8)+6;
static        bank2 bit  WPUB7                @ ((unsigned)&WPUB*8)+7;

/* Definitions for IOCB register */
static        bank2 bit  IOCB4        @ ((unsigned)&IOCB*8)+4;
static        bank2 bit  IOCB5        @ ((unsigned)&IOCB*8)+5;
static        bank2 bit  IOCB6        @ ((unsigned)&IOCB*8)+6;
static        bank2 bit  IOCB7        @ ((unsigned)&IOCB*8)+7;

/* Definitions for VRCON register */
static        bank2 bit  VR0          @ ((unsigned)&VRCON*8)+0;
static        bank2 bit  VR1          @ ((unsigned)&VRCON*8)+1;
static        bank2 bit  VR2          @ ((unsigned)&VRCON*8)+2;
static        bank2 bit  VR3          @ ((unsigned)&VRCON*8)+3;
static        bank2 bit  VP6EN             @ ((unsigned)&VRCON*8)+4;
static        bank2 bit  VRR          @ ((unsigned)&VRCON*8)+5;
static        bank2 bit  C2VREN            @ ((unsigned)&VRCON*8)+6;
static        bank2 bit  C1VREN            @ ((unsigned)&VRCON*8)+7;
```

```
/* Definitions for CM1CON0 register */
static        bank2 bit  C1CH0              @ ((unsigned)&CM1CON0*8)+0;
static        bank2 bit  C1CH1              @ ((unsigned)&CM1CON0*8)+1;
static        bank2 bit  C1R          @ ((unsigned)&CM1CON0*8)+2;
static        bank2 bit  C1POL              @ ((unsigned)&CM1CON0*8)+4;
static        bank2 bit  C1OE         @ ((unsigned)&CM1CON0*8)+5;
static volatile bank2 bit       C1OUT              @ ((unsigned)&CM1CON0*8)+6;
static        bank2 bit  C1ON         @ ((unsigned)&CM1CON0*8)+7;

/* Definitions for CM2CON0 register */
static        bank2 bit  C2CH0              @ ((unsigned)&CM2CON0*8)+0;
static        bank2 bit  C2CH1              @ ((unsigned)&CM2CON0*8)+1;
static        bank2 bit  C2R          @ ((unsigned)&CM2CON0*8)+2;
static        bank2 bit  C2POL              @ ((unsigned)&CM2CON0*8)+4;
static        bank2 bit  C2OE         @ ((unsigned)&CM2CON0*8)+5;
static volatile bank2 bit       C2OUT              @ ((unsigned)&CM2CON0*8)+6;
static        bank2 bit  C2ON         @ ((unsigned)&CM2CON0*8)+7;

/* Definitions for CM2CON1 register */
static        bank2 bit  C2SYNC             @ ((unsigned)&CM2CON1*8)+0;
static        bank2 bit  T1GSS        @ ((unsigned)&CM2CON1*8)+1;
static volatile bank2 bit       MC2OUT             @ ((unsigned)&CM2CON1*8)+6;
static volatile bank2 bit       MC1OUT             @ ((unsigned)&CM2CON1*8)+7;

/* Definitions for ANSEL register */
static        bank2 bit  ANS0         @ ((unsigned)&ANSEL*8)+0;
static        bank2 bit  ANS1         @ ((unsigned)&ANSEL*8)+1;
static        bank2 bit  ANS2         @ ((unsigned)&ANSEL*8)+2;
static        bank2 bit  ANS3         @ ((unsigned)&ANSEL*8)+3;
static        bank2 bit  ANS4         @ ((unsigned)&ANSEL*8)+4;
static        bank2 bit  ANS5         @ ((unsigned)&ANSEL*8)+5;
static        bank2 bit  ANS6         @ ((unsigned)&ANSEL*8)+6;
static        bank2 bit  ANS7         @ ((unsigned)&ANSEL*8)+7;

#if defined(_16F677) || defined(_16F685) || defined(_16F687) || defined(_16F689) ||
defined(_16F690)
/* Definitions for ANSELH register */
static        bank2 bit  ANS8         @ ((unsigned)&ANSELH*8)+0;
static        bank2 bit  ANS9         @ ((unsigned)&ANSELH*8)+1;
static        bank2 bit  ANS10              @ ((unsigned)&ANSELH*8)+2;
static        bank2 bit  ANS11              @ ((unsigned)&ANSELH*8)+3;
#endif

/* Definitions for EECON1 register */
```

```
static volatile bank3 bit       RD              @ ((unsigned)&EECON1*8)+0;
static volatile bank3 bit       WR              @ ((unsigned)&EECON1*8)+1;
static          bank3 bit  WREN         @ ((unsigned)&EECON1*8)+2;
static volatile bank3 bit       WRERR           @ ((unsigned)&EECON1*8)+3;
#if !defined(_16F687)
static          bank3 bit  EEPGD                @ ((unsigned)&EECON1*8)+7;
#endif

#if defined(_16F685) || defined(_16F690)
/* Definitions for PSTRCON register */
static          bank3 bit  STRA         @ ((unsigned)&PSTRCON*8)+0;
static          bank3 bit  STRB         @ ((unsigned)&PSTRCON*8)+1;
static          bank3 bit  STRC         @ ((unsigned)&PSTRCON*8)+2;
static          bank3 bit  STRD         @ ((unsigned)&PSTRCON*8)+3;
static          bank3 bit  STRSYNC              @ ((unsigned)&PSTRCON*8)+4;
#endif

/* Definitions for SRCON register */
static volatile bank3 bit       PULSR           @ ((unsigned)&SRCON*8)+2;
static volatile bank3 bit       PULSS           @ ((unsigned)&SRCON*8)+3;
static          bank3 bit  C2REN                @ ((unsigned)&SRCON*8)+4;
static          bank3 bit  C1SEN                @ ((unsigned)&SRCON*8)+5;
static          bank3 bit  SR0          @ ((unsigned)&SRCON*8)+6;
static          bank3 bit  SR1          @ ((unsigned)&SRCON*8)+7;

// Configuration Mask Definitions
#define CONFIG_ADDR     0x2007
// Oscillator
#define EXTCLK          0x3FFF          // External RC Clockout
#define EXTIO           0x3FFE          // External RC No Clock
#define INTCLK          0x3FFD          // Internal RC Clockout
#define INTIO       0x3FFC      // Internal RC No Clock
#define EC          0x3FFB      // EC
#define HS          0x3FFA      // HS
#define XT          0x3FF9      // XT
#define LP          0x3FF8      // LP
// Watchdog Timer
#define WDTEN           0x3FFF          // On
#define WDTDIS          0x3FF7          // Off
// Power Up Timer
#define PWRTDIS         0x3FFF          // Off
#define PWRTEN          0x3FEF          // On
// Master Clear Enable
#define MCLREN          0x3FFF          // MCLR function is enabled
```

```
#define MCLRDIS          0x3FDF       // MCLR functions as IO
// Code Protect
#define UNPROTECT        0x3FFF       // Code is not protected
#define CP          0x3FBF       // Code is protected
// Data EE Read Protect
#define UNPROTECT        0x3FFF       // Do not read protect EEPROM data
#define CPD         0x3F7F       // Read protect EEPROM data
// Brown Out Detect
#define BORDIS           0x3CFF       // BOD and SBOREN disabled
#define SWBOREN          0x3DFF       // SBOREN controls BOR function
(Software control)
#define BORXSLP          0x3EFF       // BOD enabled in run, disabled in sleep,
SBOREN disabled
#define BOREN            0x3FFF       // BOD Enabled, SBOREN Disabled
// Internal External Switch Over Mode
#define IESOEN           0x3FFF       // Enabled
#define IESODIS          0x3BFF       // Disabled
// Monitor Clock Fail-safe
#define FCMEN            0x3FFF       // Enabled
#define FCMDIS           0x37FF       // Disabled

#endif
```

Index

Made in the USA
San Bernardino, CA
12 July 2017